I0816757

THE MUSHROOM COLOR ATLAS

BOLETES
PUFF BALLS
FALSE GILLS
LINEN
SILK
WOOL
GILLED
TOOTH FUNGI
POLYPORES

THE MUSHROOM COLOR ATLAS

A Guide to Dyes and Pigments Made from Fungi

JULIE BEELER

Illustrations by YULI GATES

CHRONICLE BOOKS
SAN FRANCISCO

Library of Congress Cataloging-in-Publication Data available.

ISBN 978-1-7972-2845-7

Manufactured in China.

Design by Wynne Au-Yeung.
Illustrations by Yuli Gates.
Color wheel design by Kristine Arth, Lobster Phone.

The taxonomy and specimen identification presented in this book at the time of publication was as accurate as possible. However, with ever-evolving new discoveries from within the fungi kingdom, species can be renamed and often reclassified taxonomically. The educational content in the book is supplemental; it is not meant to replace proper training and experience in working with and identifying mushrooms. Caution is recommended when using the information provided in this book and working with mushrooms at home. The author and publisher assume no responsibility or liability for any potential damages, accidents, injuries, or loss, nor any legal, incidental, or consequential damages incurred by the reader.

10 9 8 7 6 5 4 3 2

Chronicle books and gifts are available at special quantity discounts to corporations, professional associations, literacy programs, and other organizations. For details and discount information, please contact our premiums department at corporatesales@chroniclebooks.com or at 1-800-759-0190.

Chronicle Books LLC
680 Second Street
San Francisco, California 94107
www.chroniclebooks.com

TO BRAD, THE LOVE OF MY LIFE

CONTENTS

INTRODUCTION

The Mushroom Color Atlas is a living study of colors derived from the pigments of fungi. It is a resource for everyone curious about mushrooms and the beautiful colors they can produce. This book is not so much a static, exhaustive guide for color creation as it is a departure point meant to inspire further explorations and discoveries. It's your invitation to share in the kaleidoscopic wonder of the fungi kingdom.

As a lifelong designer, artist, and educator, I have been working with color and textiles since I was a kid. When a midlife career pivot afforded me the opportunity to study the fungi kingdom, I leapt at the chance. I spent countless hours in my local forests learning to identify the mushrooms I found. I became versed in mushroom colorants and tested different methods to imbue textiles with their hues. I drew on my passion for educational interactive media and created the Mushroom Color Atlas website (mushroomcoloratlas.com) as an engaging and intuitive guide. The book in your hands is a companion to that site—an in-depth, material expansion of my experiences creating mushroom-based dyes and pigments.

The colors presented in *The Mushroom Color Atlas* are only a small slice of the countless hues that can be produced in collaboration with mushrooms. I have organized this book into three main sections: color families, mushroom types, and details about the process for coaxing color from fungi. In an era when we are increasingly reliant on technology and corporate consumer goods, creating colorants from mushrooms feels like a radical act. Working hands-on with the fruits of the natural world is an opportunity to reorient ourselves toward renewable sources and sustainable practices.

Many factors influence the chromatic qualities and outcomes of mushroom colorants. The fiber type will produce different hues, a mordant will change the dye colorant of the fabric, and a modifier will shift the color. For each color study in this book, I followed these steps:

- Foraging, gathering, and preparing the dye mushrooms
- Sourcing, scouring, and mordanting the fibers
- Creating dye baths and coaxing color from the mushrooms
- Dyeing the fibers in dye baths
- Laking the dye baths to transform them into pigments
- Making watercolor paint from the pigments
- Using the watercolor to paint swatches

A BRIEF HISTORY

It wasn't until 1969 that fungi were taxonomically separated from plants and recognized as inhabiting their own kingdom. Mushrooms are closer to animals—including humans—than they are to plants. As natural recyclers of organic matter, they have long provided great nutritional and medicinal value, but many people are astonished to find that mushrooms can also create a wide array of colors. There is documentation of dyeing fibers with mushrooms dating back to the fifteenth century, and possibly earlier. We know that polypore mushrooms were used across Africa, Asia, Europe, and the Northwest Coast of North America as a source of red, orange, yellow, gray, and purple colorants.[1] It seems certain that fungi were used at an industrial scale for a few centuries in European textile markets.[2] Books dating to the fifteenth and sixteenth centuries, held within the library at the Royal Botanic Gardens, Kew (just outside of London), contain

1. Dominique Cardon, *Natural Dyes: Sources, Tradition, Technology and Science* (London: Archetype Publications, 2007), 525.
2. Riikka Räisänen, Anja Primetta, Kirsi Niinimäki, *Dyes from Nature* (London: Archetype Publications, 2016), 21.

scientific illustrations of various dye mushrooms, though the identification of fungi in these historical texts is hindered by the fact that past knowledge of fungi species was limited. Kew's Fungarium contains specimens of dried dye mushrooms dating as far back as the 1840s. When I visited the Fungarium in 2023, I was shocked to find particles of pigment still visible around many of these 180-year-old specimens. In some cases, the paper to which these mushrooms are affixed have been inadvertently stained a range of reds, oranges, and yellows from contact with the mushroom pigments.

Miriam Rice created the modern-day movement of working with mushrooms for color in the late 1960s. While teaching natural dyeing in her children's art classes in Mendocino, California, Rice became curious about a bright yellow mushroom she had found (*Hypholoma fasciculare*, commonly known as sulfur tufts) and tossed a specimen into the dye pot with a bit of wool. Fortunately for all of us, her act resulted in a clear, bright lemon-yellow dye, inspiring her lifelong journey into extracting color from mushrooms. Rice documented and published her discoveries, which led to collaborations with Dorothy Beebee, Carla and Erik Sundstrom, and many other important mushroom dyers, researchers, and scientists. In 1980, they gathered at the first International Fungi & Fibre Symposium (an event which continues to this day) to experiment with pigment-bearing mushrooms. Rice and Beebee published important books on the subject, including *Mushrooms for Dyes, Paper, Pigments & Myco-Stix*.

Recently, there has been a surge in the research of fungi colorants and their application to textiles. More scientists from diverse backgrounds are now researching fungi, and so, as the study of mycology continues to develop with the discovery of new species—especially with the advancement of DNA technology and improved methods of cultivating mushrooms—there is sure to be an explosion of possibilities, processes, and colorants from the fungi kingdom. Ventures into commercial production of mushrooms for pigments are just starting to take shape. In the broader fungi kingdom, scientists are showing interest in several species of fungal microorganisms that can be grown to create "colorant factories" for the textile and food industries.[3] The fungi kingdom can be our powerful partner in helping to transform the dye, textile, and fashion industry, which is responsible for around 10% of global carbon dioxide emissions and polluting around 20% of global clean water.[4] These developments may help reduce the amount of industrial, synthetic colorant produced, which would contribute to a less toxic environment for us and our planet.

I hope that through *The Mushroom Color Atlas* readers will be inspired to learn more about the mycological world and come to appreciate the importance of the fungal networks and symbiotic relationships that underpin our forests. I want to draw attention to the human impact on these delicate networks and to our role as stewards of the land, with the goal of bringing positive change to environments around the globe.

---

3. Räisänen, Primetta, Niinimäki, *Dyes from Nature*, 200.
4. Walter Leal Filho, Patsy Perry, Hilde Heim, Maria Alzira Pimenta Dinis, Haruna Moda, Eromose Ebhuoma, Arminda Paco, "An overview of the contribution of the textiles sector to climate change," *Frontiers in Environmental Science* 10 (September 2022): https://www.frontiersin.org/articles/10.3389/fenvs.2022.973102.

SECTION I

COLORS

The typical human eye can perceive millions of different colors. When we see the light reflecting off a distant mauve mountainside or study a yellow-orange chanterelle mushroom up close, a cascade of chemical and electrical impulses erupts in our eyes and brains. Cone cells in our retinas hold pigments that absorb a range of visible light. Our ability to perceive color depends on the presence of colored pigments in our eyes: In order to see color we must contain color. Neurons in the retina send electrical signals up the optic nerve to the same part of the brain we use for identifying faces and reading words. The brain translates these electrical signals into all the colors we can see.

Humans love color. For millennia, we have been extracting a rainbow of colors from rocks, minerals, plants, insects, lichens, and mushrooms. The pigments and dyes created from these natural materials have been used to color everything from cave paintings to body art to clothing and adornments. We use colors to express ourselves, to signal our uniqueness and our belonging, to show who we are and what we believe.

Colors speak to our emotions, too. Our language is peppered with color-based mood metaphors: We say we are feeling "blue," or "green with envy," or are wearing "rose-colored glasses." Viewing certain colors can change our mood and even our heart rate. Immersing ourselves in the emerald green of a forest lends us feelings of optimism and well-being. Staring at turquoise ocean waters makes us feel calm. It's no wonder color feels so important to us: Color discernment has been crucial to our survival as a species. Color acuity led early humans to the ripest and most nutritious fruit, it helped us stalk and identify game, and it was critical for selecting healthy mates.

Over time, the ways we use and think of color have grown more sophisticated, as have the ways we extract and create color from natural and synthetic sources. Humans have coaxed colorants from fungi since at least the Middle Ages.[1] The fleshy forest organisms we call mushrooms are the fruiting bodies (the reproductive spore-dispersal organs) of subterranean fungi. Imagine walking through a hushed valley of evergreen trees. The air smells piney. Sunlight shines on teal-toned spruces, and hemlocks glow lime green. The ground is soft from millennia of needle-fall. If we could travel beneath the surface of the soil, we would find a complex web of fungi connecting the trees in a pulsing network of life, miles and miles of mycelium threads sharing nutrients, sugar, and water with other plants and fungi. The long, white branching filaments known as mycelium, the underground bodies of the fungus, are an ecological connective tissue, the living seam by which much of the world is stitched into relation.[2]

MUSHROOMS AND THEIR COLOR PIGMENT COMPOUNDS

Mushrooms are masters of breaking down organic matter and incorporating trace elements into their flesh. Some species of fungi alchemize these hidden wisps of dead matter into vivid colors. The kingdom of fungi comprises millions of different species, and while not all mushrooms produce colorful pigments, the ones that do can create a wide spectrum of hues. The subgenus of mushrooms *Cortinarius dermocybe* contains anthraquinones that produce shades of red. Some mushrooms host emodin,

---

1. Cardon, *Natural Dyes*, 525.
2. Merlin Sheldrake, *Entangled Life: How Fungi Make Our Worlds, Change Our Minds & Shape Our Futures* (New York: Random House, 2020), 46.

a widespread anthraquinone that creates shades of orange. White rot mushrooms, such as sulfur tufts (*Hypholoma fasciculare*), contain yellow pigments. Tooth fungi, specifically the *Hydnellum* and *Sarcodon* genuses, are full of terphenylquinones (pigments found only in mushrooms!) that produce green, blue, and purple hues. The incredible colors held in mushroom pigments mirror the vast array of hues we can encounter in the natural world.

ORGANIZING THE RAINBOW IN YOUR HANDS

The three fundamental principles for organizing color are hue, value, and saturation. *Hue* is the name assigned to a color, the property of a particular color that distinguishes it from others. *Value* refers to the amount of lightness or darkness in a color. *Saturation* refers to the degree of color intensity, the purity or vividness of a color. Each swatch in the online Mushroom Color Atlas was assigned a numeric value in a 360-degree spectrum, creating a chromatic rainbow. The selection of colors from the online atlas presented within this book is inspired by the spectrum of a rainbow. Each color family—red, orange, yellow, green, blue, and purple—has its own dedicated chapter. The hues are organized in this fashion as an homage to *The Rainbow Beneath My Feet*, the mushroom dyer's field guide written by Arleen and Alan Bessette. The introductions for each color chapter in this book draw inspiration from work dating back to the late 1700s by Abraham Gottlob Werner, a German geologist who organized a color classification system for minerals based on their external properties, and also from artist and botanical illustrator Patrick Syme, who expanded this nomenclature in the early 1800s to include the color properties of animals and vegetables.

WORKING WITH MUSHROOMS FOR COLOR

Creating dyes and pigments from mushrooms yields a quantity and quality of colors that stand in stark contrast to the hues made from synthetic dyes. The colors produced from mushrooms seem to share a visual lineage—each hue transitions smoothly to the next in a sort of familial harmony. Synthetic dyes tend to lack the emotive resonance found in natural dyes, lending them an inanimate quality. Consider for example the red color of the fly agaric mushroom (*Amanita muscaria*)—popularized by *Alice's Adventures in Wonderland* for its magical ability to make Alice change size—versus the industrial red color of a stop sign. The red of the fly agaric lives and breathes in symbiosis with its shifting environment while the color of the stop sign remains uniform, lifeless, fixed. Working with natural dyes is a time-honored, hands-on, rich experience. Learning to extract these natural colors—which requires making countless decisions with our eyes, hands, and aesthetic intuitions—gives us an intimate connection to the hues and their mushroom sources. Any artworks or crafts we produce with the natural pigments we have created are imbued with the experience of that connection.

Amateur mycologist and twentieth-century American composer John Cage once remarked, "I have come to the conclusion that much can be learned about music by devoting oneself to the mushroom." The chromatic interplay of mushroom hues, when combined, resembles the blending of individual musical tones into rich harmonies. Like music, color is a great catalyst for feeling and transformation. *The Mushroom Color Atlas* endeavors to empower makers to develop color palettes with personal meaning and resonance.

CHAPTER ONE

Red

Ruby. Crimson. Scarlet. Vermilion. Pink. Rose. Carmine. Rouge.

Red stands out. This bold primary hue anchors one end of the color spectrum with the longest wavelength of any visible color, around 700 nanometers. No other colors can be combined to create the color red. It is primal, the color of our blood. We associate red with love and anger, heat and danger. Red has no problem holding these contradictions. It can mean STOP, lend a sense of speed and power to a sports car, or signal youth and vitality in the lipstick and rouge we wear. Many food products (think sports drinks, breakfast cereals, candies) are colored red with questionable synthetic dyes (like red dye 40) to increase sales. Red's long wavelength means it is also the first color to disappear in low-light situations such as underwater or at night. But in the light of day, we are drawn to the flash of red. In the natural world, red often acts as a lure and an indicator. Flowers like tulips and poppies shine red to attract pollinators. Apples, strawberries, pomegranates, and cherries broadcast their ripeness with bright red, attracting animals that spread their seeds. Birds like the red-winged blackbird and the northern cardinal flaunt red plumage to both attract mates and establish dominance. In the forest understory, mushrooms with red pigments beam with vivacity, offering a depth of color unmatched by flat synthetic versions of red. The red hues derived from mushrooms have *charisma*—their dynamic color range extends from brash crimson to delicate pink.

The word *red* can be traced back to the proto-Indo-European word *reudh*, meaning "ruddy." Humankind has utilized red pigments for many thousands of years. Archeological sites indicate that humans were working with ochre (in the form of iron-rich stones ground into pigment) more than 300,000 years ago.[1] Our earliest forms of art and symbolism were rendered in red. Cave paintings on almost every continent attest to our enduring relationship with this mineral hue. A spectrum of red minerals such as ruby, garnet, and cinnabar (responsible for the brilliant shade of vermilion) are found in landscapes around the world.

Before modern times, red dyes primarily came from a tiny, pinhead-size female insect known as a cochineal (*Dactylopius coccus*) or the ruby-red roots of a flowering plant called madder (*Rubia tinctorum*). In ancient Mesoamerica, cochineal bugs were scraped from the paddles of the prickly pear cactus and used as a source of dye and pigment. Today, we can still find cochineal coloring in our food and cosmetic products. From the sixteenth to nineteenth centuries, madder root proved a less expensive alternative to cochineal and was used to dye such emblematic garb as the British military's famous red coats. Even Tutankhamen's tomb (completed circa 1324 BCE) held madder dyes. In the eighteenth and nineteenth centuries, madder was used to make the highly coveted Turkey red, which, through a secretly held dye process, achieved the era's most brilliant carmine hue.

---

1. Gemma Tarlach, "What the Ancient Pigment Ochre Tells Us About the Human Mind," *Discover*, March 15, 2018, https://www.discovermagazine.com/planet-earth/prehistoric-use-of-ochre-can-tell-us-about-the-evolution-of-humans.

Mushrooms contain a few different red pigment compounds known as anthraquinones, the same pigment-based organic compounds found in madder (alizarin) and cochineal (carminic acid). A subgenus of mushrooms, *Cortinarius dermocybe*, has aptly named dermocybin and dermorubin pigments. Dermocybin is responsible for the brilliant red pigment and dye from *C. neosanguineus*. As a pigment, it creates a striking red that can be modified to an intense orange-red, a stately plum red, or a dark, purplish red. Dermorubin is responsible for the rose-pink colorants found in *C. ominosus* and *Hypomyces lactifluorum* and creates a variety of charming hues. Red colorants are often achieved through modification of an alkaline (pH9) dye bath and improved with mordants such as tin, titanium oxalate, alum, and aluminum acetate with tannins in combination with cellulose fibers. Reddish browns can be extracted from a variety of mushrooms either as a dye or a pigment. The tooth fungus *Echinodontium tinctorium* produces an intense burgundy red to dusty brown-red in an alkaline (pH9) environment, while *Pisolithus arhizus*, a puffball mushroom, offers reds and browns subtly tinged with gold, giving a rust-like appearance to the pigment.

MUSHROOMS FEATURED IN THIS SECTION:

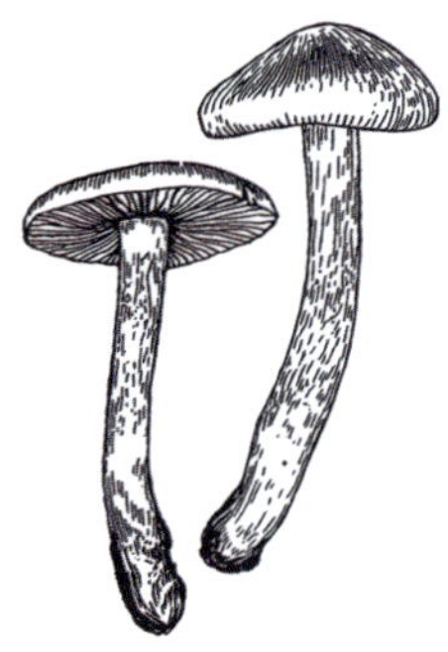

Cortinarius neosanguineus

Cortinarius ominosus

Cortinarius subcroceofolius

Cortinarius uliginosus

Echinodontium tinctorium

Hypomyces lactifluorum

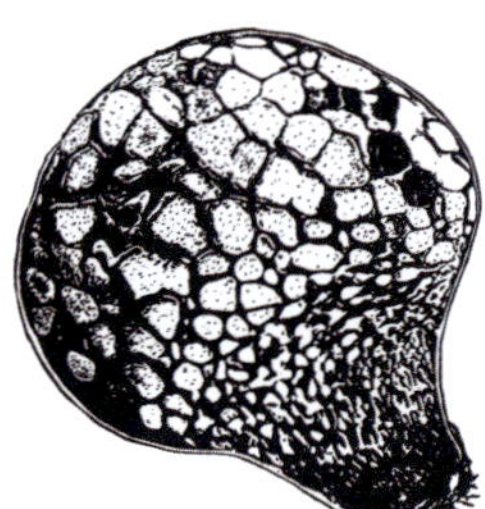

Pisolithus arhizus

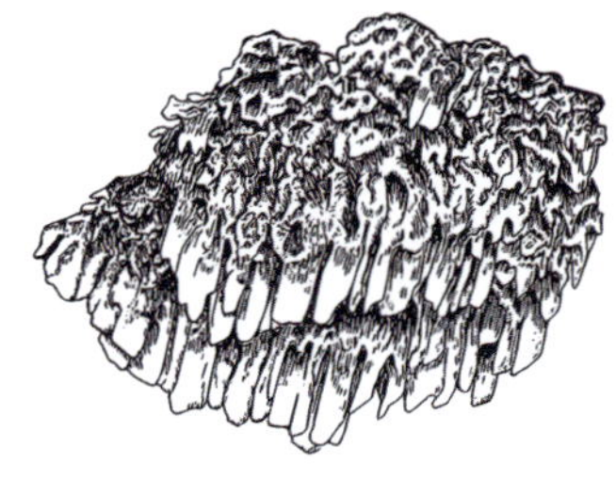

Pycnoporellus alboluteus

1. *Cortinarius ominosus*, Silk, Tin
2. *Cortinarius ominosus*, Silk, Alum
3. *Cortinarius ominosus*, Linen, Titanium Oxalate
4. *Echinodontium tinctorium*, Linen, Aluminum Acetate, pH10
5. *Cortinarius neosanguineus*, Pigment, Copper Acetate
6. *Cortinarius ominosus*, Wool, Alum
7. *Hypomyces lactifluorum*, Silk, Alum, pH10
8. *Hypomyces lactifluorum*, Silk, Tin, pH10
9. *Cortinarius ominosus*, Wool, Tin

10. *Cortinarius neosanguineus*, Pigment, Citric Acid
11. *Hypomyces lactifluorum*, Pigment, Base (Fresh)
12. *Cortinarius neosanguineus*, Pigment, Citric Acid, pH10
13. *Cortinarius ominosus*, Wool, Tin, pH9
14. *Cortinarius neosanguineus*, Wool, Alum, pH10
15. *Cortinarius neosanguineus*, Silk, Alum, pH10
16. *Hypomyces lactifluorum*, Pigment, Copper Acetate (Fresh)
17. *Cortinarius neosanguineus*, Silk, Tin, pH10
18. *Cortinarius neosanguineus*, Linen, Aluminum Acetate, pH10

19. *Hypomyces lactifluorum*, Wool, Tin, pH10
20. *Cortinarius neosanguineus*, Pigment, Soda Ash
21. *Cortinarius neosanguineus*, Pigment, Alum
22. *Cortinarius neosanguineus*, Pigment, Base
23. *Hypomyces lactifluorum*, Pigment, Soda Ash (Fresh)
24. *Hypomyces lactifluorum*, Wool, Alum, pH10
25. *Hypomyces lactifluorum*, Wool, Alum, pH9 (Fresh)
26. *Cortinarius neosanguineus*, Linen, Titanium Oxalate, pH10
27. *Cortinarius neosanguineus*, Pigment, Alum, pH10

28. *Hypomyces lactifluorum*, Wool, Tin, pH9 (Fresh)
29. *Cortinarius neosanguineus*, Pigment, Soda Ash, pH10
30. *Cortinarius neosanguineus*, Pigment, Base, pH10
31. *Cortinarius neosanguineus*, Linen, Titanium Oxalate
32. *Hypomyces lactifluorum*, Wool, Iron, pH10
33. *Cortinarius neosanguineus*, Pigment, Copper Acetate, pH10
34. *Cortinarius subcroceofolius*, Pigment, Alum
35. *Cortinarius subcroceofolius*, Pigment, Base
36. *Cortinarius neosanguineus*, Linen, Aluminum Acetate

37. *Echinodontium tinctorium*, Pigment, Alum
38. *Cortinarius ominosus*, Silk, Tin, pH9
39. *Hypomyces lactifluorum*, Pigment, Base
40. *Hypomyces lactifluorum*, Pigment, Soda Ash
41. *Pycnoporellus alboluteus*, Pigment, Soda Ash
42. *Cortinarius uliginosus*, Pigment, Soda Ash
43. *Cortinarius uliginosus*, Pigment, Citric Acid
44. *Cortinarius uliginosus*, Pigment, Base
45. *Cortinarius uliginosus*, Pigment, Alum

46. *Cortinarius ominosus*, Silk, Alum, pH9
47. *Cortinarius ominosus*, Wool, Alum, pH9
48. *Cortinarius ominosus*, Linen, Aluminum Acetate, pH9
49. *Hypomyces lactifluorum*, Pigment, Copper Acetate (Fresh)
50. *Hypomyces lactifluorum*, Pigment, Copper Acetate
51. *Cortinarius subcroceofolius*, Pigment, Soda Ash
52. *Cortinarius ominosus*, Pigment, Citric Acid, pH9
53. *Cortinarius ominosus*, Pigment, Soda Ash, pH9
54. *Cortinarius ominosus*, Pigment, Alum, pH9

55. *Cortinarius ominosus*, Pigment, Alum
56. *Cortinarius ominosus*, Pigment, Soda Ash
57. *Echinodontium tinctorium*, Pigment, Soda Ash
58. *Cortinarius ominosus*, Pigment, Base
59. *Cortinarius subcroceofolius*, Pigment, Alum, pH5
60. *Echinodontium tinctorium*, Pigment, Alum, pH10
61. *Echinodontium tinctorium*, Pigment, Citric Acid, pH10
62. *Pisolithus arhizus*, Pigment, Alum
63. *Cortinarius subcroceofolius*, Pigment, Base, pH5

64. *Cortinarius ominosus*, Pigment, Base, pH9
65. *Cortinarius subcroceofolius*, Pigment, Soda Ash, pH5
66. *Cortinarius ominosus*, Pigment, Iron, pH9
67. *Pisolithus arhizus*, Pigment, Soda Ash
68. *Cortinarius ominosus*, Pigment, Copper Acetate, pH9
69. *Cortinarius ominosus*, Pigment, Copper Acetate
70. *Echinodontium tinctorium*, Pigment, Copper Acetate, pH10
71. *Echinodontium tinctorium*, Pigment, Base, pH10
72. *Cortinarius neosanguineus*, Pigment, Iron, pH10

CHAPTER TWO

Orange

Amber. Saffron. Ochre. Tangerine. Ginger. Marigold. Sienna.

Orange lives between red and yellow on the color spectrum, reflecting light with a wavelength of 600 nanometers. Orange catches our eye and demands our attention. Fluorescent synthetic orange is often employed to convey immediate visibility on safety items such as traffic cones, personal flotation devices, and even astronaut suits. We associate orange with warmth, wisdom, and energy. The orange hues derived from mushrooms underscore these associations by possessing a more grounded quality than their synthetic counterparts. Thanks to overlapping red and yellow colorants in fungi, orange mushroom dyes and pigments produce warm, earthy hues.

In contemporary Western culture, we tend to associate orange with autumn, excitement, visibility, and warning, but humans have long assigned religious and spiritual meaning to the color. In Confucianism, orange is the color of transformation as symbolized by sunrise and sunset. Fire—perhaps the ultimate agent of change—burns orange, and the ancient Greeks often depicted their god of fire, Prometheus, clad in orange finery. Likewise, the orange feathers of the mythical phoenix signal a fiery rebirth. The tones of the saffron-colored robes worn by Buddhist monks and Hindu swamis resemble some of the orange hues created from mushrooms.

The English etymology of the word *orange* can be traced back to the Old French *pomme d'orenge* (the fruit of the orange tree), which is derived from the Arabic word *nāranj*. Expressed somewhat sparingly in the natural world, vibrant orange hues can be found in minerals, including native copper and amber (one of the few organic gemstones). Most commonly, we find natural displays of orange in autumn leaves, roadside tiger lilies, and in the ripe fruit of its citrus namesake. Orange-colored animals and insects elicit emotions ranging from apprehension (a prowling tiger) to wonder and delight (a monarch butterfly). In foods such as carrots, mangoes, squash, and sweet potatoes, orange signals the presence of the antioxidant beta-carotene, which supports healthy skin, hair, and eyes.

Mushroom-derived orange hues are more prevalent and easier to create than pure reds or yellows. A few factors contribute to this, including the ability to isolate specific pigment chemical compounds during the dyeing process. Another factor is the age at which the mushroom is harvested. For example, when harvesting some species of *Cortinarius dermocybe*, the younger specimens offer more yellowish-orange colors, whereas older, more mature specimens give dark red and reddish-brown colors.[1] Different chemical compounds, appearing sometimes singly and other times in combination, contribute to the orange colorants found in mushrooms. Knowing the names and properties of these pigments can help guide you, especially as you test the dye potential of newly found mushrooms.

---

1. Thomas Bechtold, Avinash P. Manian, Tung Pham, *Handbook of Natural Colorants*, 2nd ed. (Hoboken, NJ: Wiley, 2023), 193.

C. dermocybe is a gilled mushroom subgenus that contains a variety of pigments, one of which is emodin. This orange pigment is the most widely distributed anthraquinone, found not just in mushrooms but also in insects and flowering plants. Other pigment compounds include parietin (historically referenced as physcion), which creates orange-yellow tones; fallacinol, which makes orange-reds; flavomentin, which creates orange-yellows; and pulvinic acid, which forms yellow and orange-red colors like those derived from fungi of the *Boletus* genus. Often a range of orange hues can be achieved by extracting mushroom pigments in an acidic (pH4) dye bath, as is the case with *Pycnoporellus alboluteus*, a polypore mushroom. Orange colorants are made more vivid and lightfast with the addition of mordants such as tin, titanium oxalate, and alum. *Echinodontium tinctorium*, a tooth fungus, produces salmony oranges in combination with alum mordants. *Pisolithus arhizus*, a puffball mushroom, and *Phaeolus schweinitzii*, a polypore, produce a golden orange and orange-brown in combination with tin. Curiously, *Hypomyces lactifluorum*, the lobster mushroom (which is actually a mushroom-parasitizing fungus), dyes silk fibers a salmon-orange color rather than the red its name might suggest. When the lobster mushroom's pigment is exposed to acidic modifiers like alum or citric acid, it will also produce a striking orange paint.

MUSHROOMS FEATURED IN THIS SECTION:

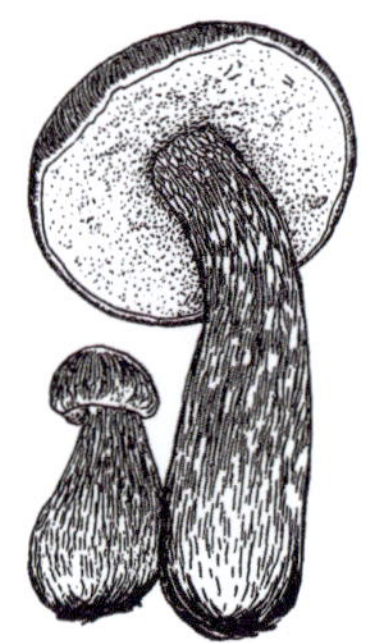

Aureoboletus mirabilis

Boletus rex-veris

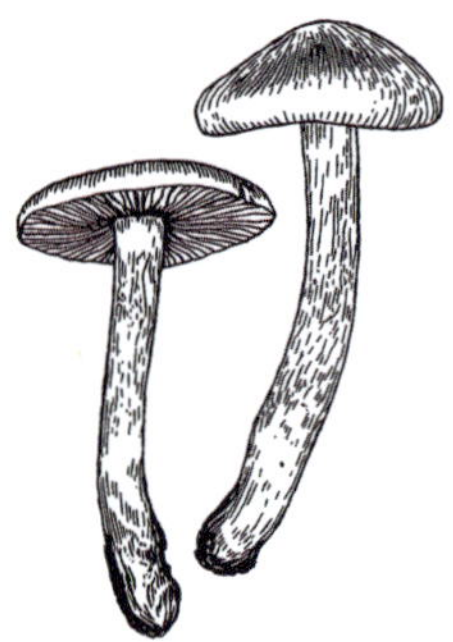

Cortinarius neosanguineus

Cortinarius ominosus

Cortinarius subcroceofolius

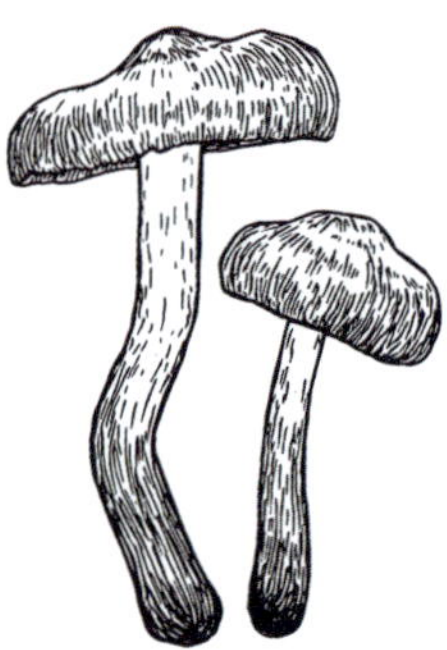

Cortinarius uliginosus

Echinodontium tinctorium

Gymnopilus ventricosus

Hypomyces lactifluorum

Inonotus obliquus

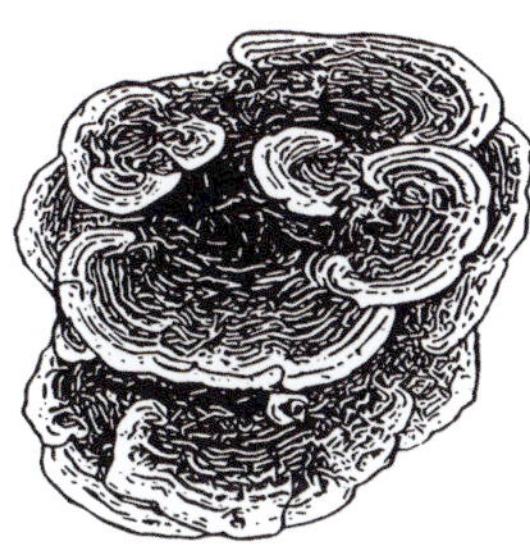

Phaeolus schweinitzii

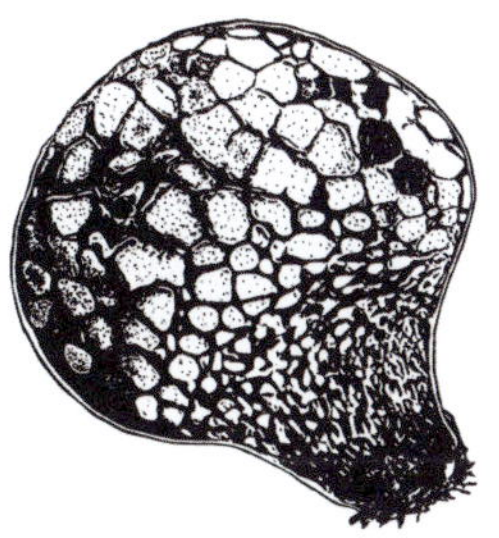

Pisolithus arhizus

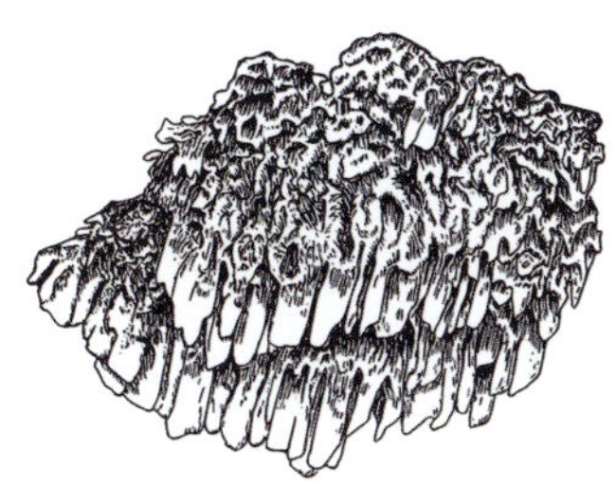

Pycnoporellus alboluteus

1. *Pycnoporellus alboluteus*, Pigment, Citric Acid
2. *Phaeolus schweinitzii*, Linen, Titanium Oxalate, pH5
3. *Echinodontium tinctorium*, Silk, Tin
4. *Inonotus obliquus*, Pigment, Iron
5. *Boletus rex-veris*, Wool, Tin, pH5 (Fresh)
6. *Boletus rex-veris*, Wool, Tin, pH5
7. *Aureoboletus mirabilis*, Wool, Tin, pH5
8. *Phaeolus schweinitzii*, Pigment, Citric Acid
9. *Boletus rex-veris*, Silk, Tin, pH5

10. *Phaeolus schweinitzii*, Wool, Tin, pH9
11. *Aureoboletus mirabilis*, Silk, Alum, pH5
12. *Pycnoporellus alboluteus*, Wool, Alum, pH4
13. *Pycnoporellus alboluteus*, Wool, Tin, pH4
14. *Pycnoporellus alboluteus*, Silk, Tin, pH4
15. *Echinodontium tinctorium*, Linen, Iron
16. *Cortinarius uliginosus*, Wool, Alum
17. *Cortinarius uliginosus*, Wool, Tin
18. *Echinodontium tinctorium*, Silk, Alum

19. *Echinodontium tinctorium*, Wool, Alum
20. *Echinodontium tinctorium*, Wool, Tin
21. *Cortinarius subcroceofolius*, Wool, Tin, pH5
22. *Cortinarius subcroceofolius*, Silk, Alum, pH5
23. *Cortinarius subcroceofolius*, Silk, Iron, pH5
24. *Cortinarius subcroceofolius*, Wool, Alum, pH5
25. *Cortinarius subcroceofolius*, Wool, Iron, pH5
26. *Phaeolus schweinitzii*, Wool, Tin, pH5
27. *Cortinarius subcroceofolius*, Silk, Tin, pH5

28. *Cortinarius subcroceofolius*, Wool, Alum
29. *Cortinarius neosanguineus*, Silk, Tin
30. *Cortinarius neosanguineus*, Pigment, Iron
31. *Hypomyces lactifluorum*, Pigment, Citric Acid (Fresh)
32. *Hypomyces lactifluorum*, Pigment, Alum (Fresh)
33. *Pycnoporellus alboluteus*, Pigment, Alum
34. *Pycnoporellus alboluteus*, Pigment, Base
35. *Cortinarius subcroceofolius*, Wool, Tin
36. *Pycnoporellus alboluteus*, Pigment, Iron

37. *Hypomyces lactifluorum*, Linen, Iron, pH10
38. *Cortinarius subcroceofolius*, Linen, Iron, pH5
39. *Echinodontium tinctorium*, Linen, Iron, pH10
40. *Echinodontium tinctorium*, Silk, Tin, pH10
41. *Cortinarius subcroceofolius*, Wool, Iron
42. *Cortinarius subcroceofolius*, Pigment, Iron
43. *Hypomyces lactifluorum*, Linen, Aluminum Acetate, pH10 (Fresh)
44. *Echinodontium tinctorium*, Silk, Alum, pH10
45. *Echinodontium tinctorium*, Wool, Alum, pH10

46. *Hypomyces lactifluorum*, Silk, Iron, pH10
47. *Hypomyces lactifluorum*, Silk, Tin, pH9 (Fresh)
48. *Hypomyces lactifluorum*, Silk, Alum, pH9 (Fresh)
49. *Hypomyces lactifluorum*, Linen, Iron, pH9 (Fresh)
50. *Echinodontium tinctorium*, Wool, Tin, pH10
51. *Hypomyces lactifluorum*, Silk, Iron, pH9 (Fresh)
52. *Hypomyces lactifluorum*, Pigment, Alum, pH10
53. *Cortinarius neosanguineus*, Wool, Tin
54. *Cortinarius neosanguineus*, Silk, Alum

55. *Echinodontium tinctorium*, Pigment, Citric Acid
56. *Cortinarius neosanguineus*, Wool, Alum
57. *Cortinarius ominosus*, Pigment, Citric Acid, pH9 (Fresh)
58. *Hypomyces lactifluorum*, Pigment, Citric Acid
59. *Hypomyces lactifluorum*, Pigment, Copper Acetate (Fresh)
60. *Hypomyces lactifluorum*, Pigment, Iron (Fresh)
61. *Cortinarius subcroceofolius*, Pigment, Citric Acid
62. *Echinodontium tinctorium*, Pigment, Base
63. *Cortinarius uliginosus*, Pigment, Iron

64. *Echinodontium tinctorium*, Pigment, Iron
65. *Hypomyces lactifluorum*, Pigment, Iron
66. *Cortinarius subcroceofolius*, Pigment, Iron, pH5
67. *Cortinarius subcroceofolius*, Pigment, Citric Acid, pH5
68. *Pisolithus arhizus*, Pigment, Base
69. *Cortinarius ominosus*, Pigment, Iron
70. *Pisolithus arhizus*, Linen, Titanium Oxalate
71. *Pisolithus arhizus*, Linen, Aluminum Acetate
72. *Pisolithus arhizus*, Wool, Tin

73. *Pisolithus arhizus*, Pigment, Citric Acid
74. *Gymnopilus ventricosus*, Pigment, Soda Ash
75. *Pisolithus arhizus*, Linen, Iron
76. *Gymnopilus ventricosus*, Pigment, Citric Acid
77. *Gymnopilus ventricosus*, Pigment, Alum
78. *Phaeolus schweinitzii*, Pigment, Soda Ash
79. *Phaeolus schweinitzii*, Pigment, Alum
80. *Phaeolus schweinitzii*, Pigment, Base
81. *Pisolithus arhizus*, Silk, Tin

82. *Pisolithus arhizus*, Silk, Alum
83. *Pisolithus arhizus*, Silk, Iron
84. *Pisolithus arhizus*, Wool, Alum

CHAPTER THREE

Yellow

Lemon. Gold. Mustard. Sunshine. Canary. Honey. Gamboge.

As one of the three primary colors (along with red and blue), yellow cannot be created by combining any other colors. It lives between green and orange on the color wheel, and its wavelength is around 570 nanometers, right in the middle of the visible spectrum. Yellow has long been associated with knowledge and wisdom, happiness and optimism, expression and warmth. The yellow sun is one of humanity's most enduring symbols, and many ancient cultures attributed power and strength to their sun deities. Yellow is complex: It can be soft and soothing, or sharp, piercing, and loud. Because of its high visibility (especially when produced as a synthetic fluorescent), yellow also denotes caution in traffic lights, police tape, and road signs.

Etymologically, gold and yellow are practically synonymous; gold's name derives from the Old English word *geolu*, meaning "yellow." Coveted throughout human history, gold's lustrous, non-tarnishing color has come to be associated with desire. Orpiment, the glittering (and deadly) yellow mineral containing 60% arsenic, was sought-after in ancient times because of its resemblance to gold. This pigment can be found in Egyptian art, illuminating the manuscripts of the ninth-century Book of Kells, and adorning the walls of the Taj Mahal.[1]

In nature, daffodils and forsythia flowers, the heralds of spring, get their exuberant hues from carotenoid pigments, the same colorants responsible for goldfinches' and warblers' yellow feathers. The carotenoid group of pigments is widespread in plants, giving corn, bananas, and lemons their signature colors. The spectrum of yellow colors derived from mushroom pigments ranges from a light, off-white beige to creamy, lemony yellow to bold, saturated golds. Some of the most common yellow mushroom pigments are tinged with smoky browns and olive greens, giving them a grounded, earthy feeling. Many mushrooms can produce beige-yellows that some dyers find rather ho-hum; instead of labeling these various species as "dye duds," they are important when considering the range of hues created from mushrooms.

Flavonoids are the primary source of yellow from most natural dyes and have recently been touted for their antioxidant role in human health (foods like lemons, onions, and white wines contain yellow flavonoids). Mushrooms, however, contain different kinds of yellow pigments: grevillines (that impart yellow color variations), methoxy-flavomannin (water soluble yellow pigment), and pulvinic acid derivatives, particularly in the *Boletus* genus. The gilled mushroom subgenus *Cortinarius dermocybe* has fifteen different anthraquinone pigments,[2] some of which are responsible for the brilliant yellows and oranges from *C. subcroceofolius*. However, when the dye is transformed into a pigment, a brick red is created rather than the yellow color one might expect. Lemony yellows can

1. Kassia St. Clair, *The Secret Lives of Color* (New York: Penguin Books, 2017), 82.
2. Räisänen, Primetta, Niinimäki, *Dyes from Nature*, 54.

be created from what are known as white rot mushrooms (such as *Gymnopilus ventricosus*). These wood-decaying fungi can be found growing on living or dead deciduous or conifer trees.

The aptly nicknamed dyer's polypore mushroom—*Phaeolus schweinitzii*—offers a substantive natural dye, meaning it doesn't require a mordant to produce a vibrant yellow, though its lightfastness will be improved with the addition of tin, titanium oxalate, alum, and aluminum acetate with tannins in combination with cellulose fibers. These mordants can help accomplish a beautiful spectrum of yellow hues while improving the dye's permanence. As for bolete mushrooms, *Aureoboletus mirabilis* produces sunshine yellow in combination with alum mordants, *Xerocomellus zelleri* makes a golden yellow in combination with tin, and *Boletus rex-veris* creates an olive-yellow in combination with iron.

To produce more saturated, vibrant yellows, make the dye bath mildly acidic (pH4) using household distilled white vinegar. Likewise, adding 1% chalk (calcium carbonate) to the dye pot before dropping in the fibers will help "bloom" the color, giving it a clearer, sharper yellow. But keep in mind that adding chalk to the dye pot will also increase the opaqueness of any pigment created through the lake-making process.

MUSHROOMS FEATURED IN THIS SECTION:

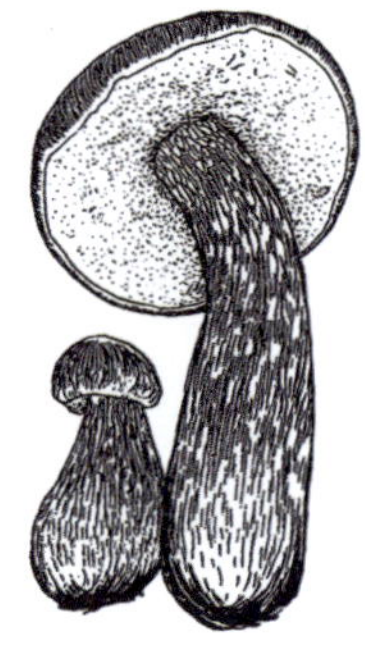

Aureoboletus mirabilis

Boletus rex-veris

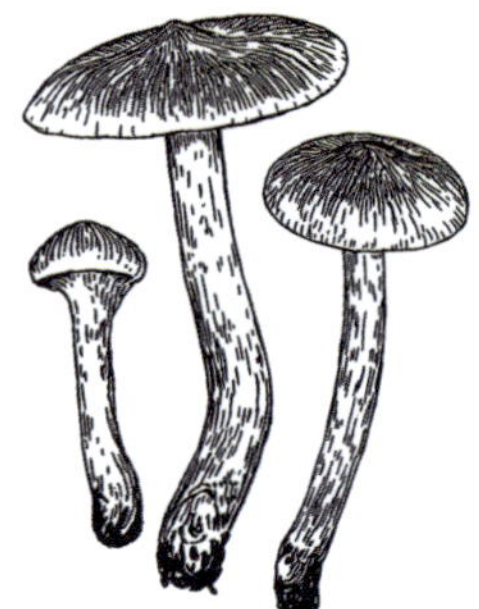

Cortinarius subcroceofolius

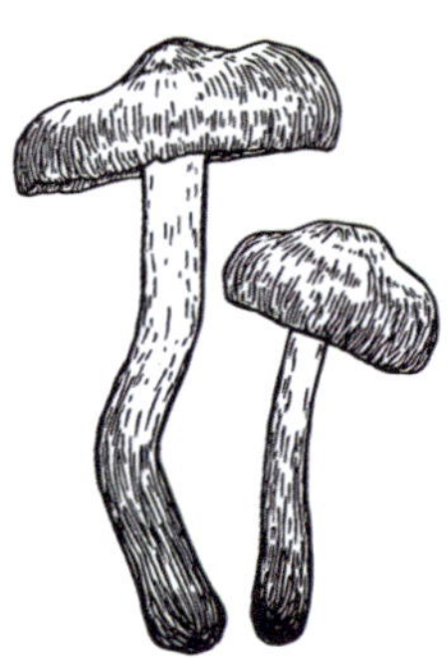

Cortinarius uliginosus

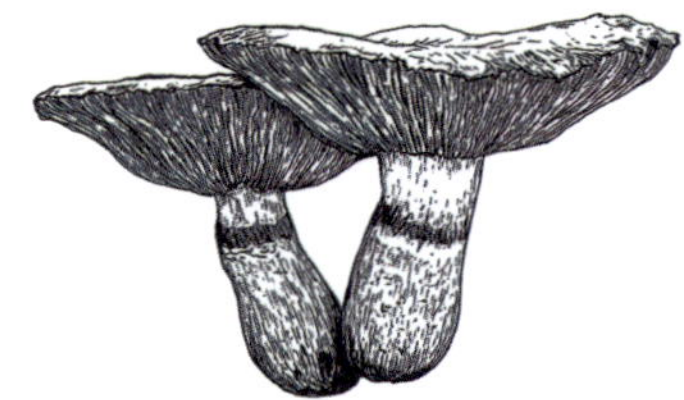

Gymnopilus ventricosus

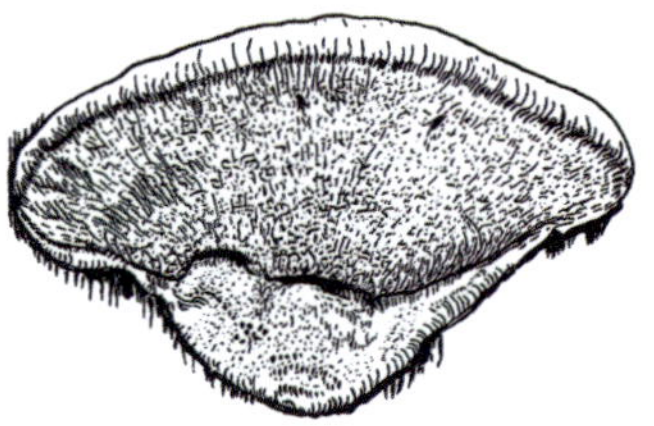

Hapalopilus nidulans

Inonotus obliquus

Omphalotus olivascens

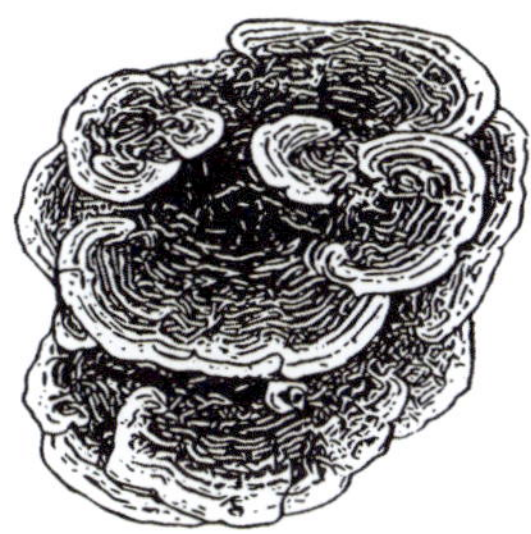

Phaeolus schweinitzii

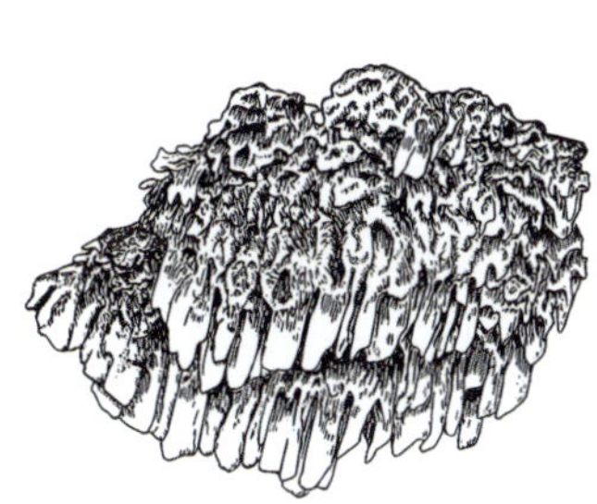

Pycnoporellus alboluteus

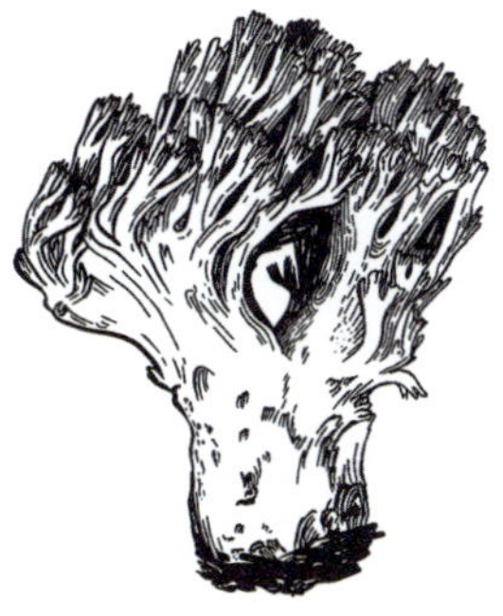

Ramaria spp.

Xerocomellus zelleri

1. *Aureoboletus mirabilis*, Linen, Titanium Oxalate, pH5
2. *Gymnopilus ventricosus*, Linen, Titanium Oxalate, pH4
3. *Boletus rex-veris*, Silk, Alum, pH5
4. *Boletus rex-veris*, Pigment, Base (Fresh)
5. *Gymnopilus ventricosus*, Silk, Alum, pH4
6. *Gymnopilus ventricosus*, Silk, Tin, pH4
7. *Xerocomellus zelleri*, Silk, Tin, pH5
8. *Phaeolus schweinitzii*, Silk, Alum, pH5
9. *Boletus rex-veris*, Wool, Alum, pH5 (Fresh)

10. *Phaeolus schweinitzii*, Silk, Alum, pH9
11. *Phaeolus schweinitzii*, Silk, Tin, pH9
12. *Boletus rex-veris*, Silk, Tin, pH5 (Fresh)
13. *Xerocomellus zelleri*, Silk, Iron, pH5
14. *Xerocomellus zelleri*, Silk, Alum, pH5
15. *Aureoboletus mirabilis*, Silk, Alum, pH5 (Fresh)
16. *Aureoboletus mirabilis*, Silk, Iron, pH5 (Fresh)
17. *Aureoboletus mirabilis*, Wool, Iron, pH5 (Fresh)
18. *Boletus rex-veris*, Wool, Iron, pH5 (Fresh)

19. *Gymnopilus ventricosus*, Linen, Aluminum Acetate, pH4
20. *Phaeolus schweinitzii*, Linen, Titanium Oxalate, pH9
21. *Xerocomellus zelleri*, Wool, Alum, pH5
22. *Ramaria* spp., Linen, Titanium Oxalate
23. *Cortinarius subcroceofolius*, Silk, Tin
24. *Gymnopilus ventricosus*, Wool, Alum, pH4
25. *Inonotus obliquus*, Silk, Tin, pH5
26. *Gymnopilus ventricosus*, Wool, Tin, pH4
27. *Aureoboletus mirabilis*, Wool, Alum, pH5 (Fresh)

28. *Boletus rex-veris*, Linen, Titanium Oxalate, pH5 (Fresh)
29. *Boletus rex-veris*, Pigment, Alum (Fresh)
30. *Aureoboletus mirabilis*, Pigment, Citric Acid
31. *Cortinarius uliginosus*, Silk, Alum
32. *Xerocomellus zelleri*, Pigment, Iron
33. *Ramaria* spp., Pigment, Iron
34. *Xerocomellus zelleri*, Wool, Tin, pH5
35. *Boletus rex-veris*, Silk, Alum, pH5 (Fresh)
36. *Phaeolus schweinitzii*, Wool, Alum, pH5

37. *Phaeolus schweinitzii*, Silk, Tin, pH5
38. *Aureoboletus mirabilis*, Silk, Tin, pH5
39. *Aureoboletus mirabilis*, Silk, Tin, pH5 (Fresh)
40. *Aureoboletus mirabilis*, Wool, Tin, pH5
41. *Cortinarius uliginosus*, Silk, Tin
42. *Ramaria* spp., Wool, Tin
43. *Inonotus obliquus*, Wool, Tin, pH5
44. *Pycnoporellus alboluteus*, Silk, Alum, pH4
45. *Phaeolus schweinitzii*, Wool, Alum, pH9

46. *Ramaria* spp., Pigment, Alum
47. *Xerocomellus zelleri*, Linen, Aluminum Acetate, pH5
48. *Aureoboletus mirabilis*, Linen, Aluminum Acetate, pH5 (Fresh)
49. *Boletus rex-veris*, Linen, Aluminum Acetate, pH5 (Fresh)
50. *Omphalotus olivascens*, Pigment, Citric Acid
51. *Omphalotus olivascens*, Pigment, Alum
52. *Inonotus obliquus*, Linen, Titanium Oxalate, pH5
53. *Aureoboletus mirabilis*, Wool, Alum, pH5
54. *Aureoboletus mirabilis*, Wool, Iron, pH5

55. *Boletus rex-veris*, Silk, Iron, pH5 (Fresh)
56. *Omphalotus olivascens*, Pigment, Base
57. *Boletus rex-veris*, Wool, Alum, pH5
58. *Xerocomellus zelleri*, Wool, Iron, pH5
59. *Phaeolus schweinitzii*, Linen, Aluminum Acetate, pH5
60. *Gymnopilus ventricosus*, Silk, Iron, pH4
61. *Ramaria* spp., Pigment, Citric Acid
62. *Ramaria* spp., Pigment, Base
63. *Inonotus obliquus*, Silk, Alum, pH5

64. *Cortinarius subcroceofolius*, Linen, Titanium Oxalate
65. *Ramaria* spp., Pigment, Soda Ash
66. *Boletus rex-veris*, Pigment, Citric Acid (Fresh)
67. *Cortinarius subcroceofolius*, Linen, Titanium Oxalate, pH5
68. *Hapalopilus nidulans*, Linen, Aluminum Acetate
69. *Aureoboletus mirabilis*, Linen, Aluminum Acetate, pH5
70. *Phaeolus schweinitzii*, Linen, Aluminum Acetate, pH9
71. *Boletus rex-veris*, Linen, Aluminum Acetate, pH5
72. *Boletus rex-veris*, Linen, Titanium Oxalate, pH5

73. *Aureoboletus mirabilis*, Pigment, Soda Ash
74. *Aureoboletus mirabilis*, Pigment, Alum
75. *Ramaria* spp., Silk, Tin
76. *Aureoboletus mirabilis*, Pigment, Iron
77. *Inonotus obliquus*, Wool, Alum, pH5
78. *Phaeolus schweinitzii*, Linen, Iron, pH9
79. *Phaeolus schweinitzii*, Pigment, Citric Acid, pH5
80. *Phaeolus schweinitzii*, Pigment, Alum, pH5
81. *Phaeolus schweinitzii*, Pigment, Soda Ash, pH5

82. *Omphalotus olivascens*, Pigment, Iron
83. *Aureoboletus mirabilis*, Silk, Iron, pH5
84. *Phaeolus schweinitzii*, Pigment, Copper Acetate, pH9

CHAPTER FOUR

Green

Emerald. Jade. Celadon. Lime. Forest. Olive. Teal. Mint. Spring.

Our eyes have adapted to the preponderance of green on our planet: One-third of the cone cells in a typical human eye are dedicated to green's medium-length waves, giving us the ability to perceive more shades of this verdant hue than any other color. Green resides near the middle of the visible spectrum with a wavelength of 490 nanometers and sits between yellow and blue on the color wheel. Green is affiliated with spring, lushness, and the natural world. The most abundant pigment on earth is the green chlorophyll inside plant cells. Since photosynthesis is the basis of our planet's food chain, it's not much of a stretch to say that green is life itself.

The word *green* can be traced to the German word *grün* and the Old English word *grene*, both of which carry connotations of grass and growth. *Viridis*, the Latin word for green, is derived from the verb *vireo*, which means "to be verdant, to sprout." But the color green calls to mind more than nature's bounty; we associate green with money, jealousy, and nausea. We might also call someone who is inexperienced with a particular task green. Lately, the word has become shorthand for *environmentally responsible*, with companies touting their green products and services.

For centuries, visual artists longed for a truly green colorant born of natural materials (like they had previously developed with the color red). Green had always been produced by mixing yellow and blue pigments. It wasn't until 1775, when Carl Wilhelm Scheele invented the first pigment out of copper arsenites, that a convincing green pigment was born. Scheele's green became a highly favored pigment used in everything from textiles to wallpaper to paint.[1]

The neon, crisp hues of synthetically produced greens do not represent the diverse range of values and subtleness found in greens derived from nature. Natural copper oxidizes from reddish orange to light green, and other green minerals, such as emerald and jade, have mesmerized humans for centuries; jade was regarded as the most precious stone in ancient China, and the first known emerald mines were in Egypt (Cleopatra was known to have a passion for the green gems).

The hues of green derived from mushrooms run the gamut from bright, bold yellow-greens to deep, dark brown-greens and everything in between. The values of these greens come from deep within the earth and spring to life through fungal pigment compounds—terphenylquinones—which are found nowhere else in nature. The blue-green derived from mushrooms can be tricky to achieve but is well worth the effort. Yellowish and brownish greens are more easily created from a variety of mushroom species in combination with a iron mordant. Getting a true emerald green from mushrooms is very challenging; most greens tend to range from brown to yellow to blue. Often these pigments require an alkaline (pH9)

---

1. St. Clair, *The Secret Lives of Color*, 224.

shift with soda ash to coax out various shades. The combination of fiber choice and proper mordant will also determine whether you get a green versus a blue color. Greens tend to gain vividness and permanence with mordants such as iron, tin, and alum.

Boletopsis grisea is a simple and forgiving mushroom to work with, and it creates a deep range of greens in an alkaline (pH9) dye bath. The *Hydnellum*, *Phellodon* and *Sarcodon* genera, known as tooth fungi, encompass a variety of species that make mesmerizing shades of blue-green, each with unique depth and richness from the various pigment compounds working together in an alkaline (pH9) dye bath. Often, members of the genus *Boletus*, such as *Aureoboletus mirabilis*, in combination with a iron mordant, will produce shades of olive green. Using copper acetate as a modifier will shift pigments to a bright, grassy green, as is the case with *Omphalotus olivascens*, and to a more emerald green with *Hydnellum caeruleum*.

MUSHROOMS FEATURED IN THIS SECTION:

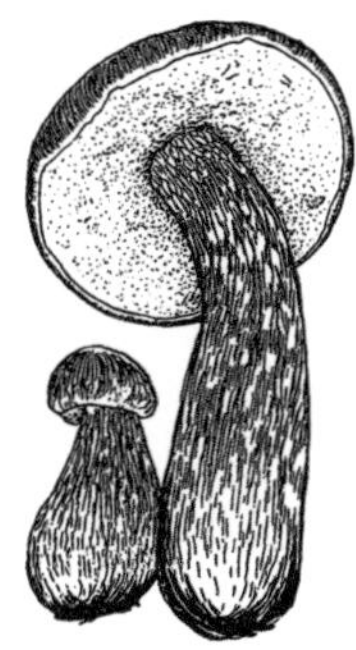

Aureoboletus mirabilis

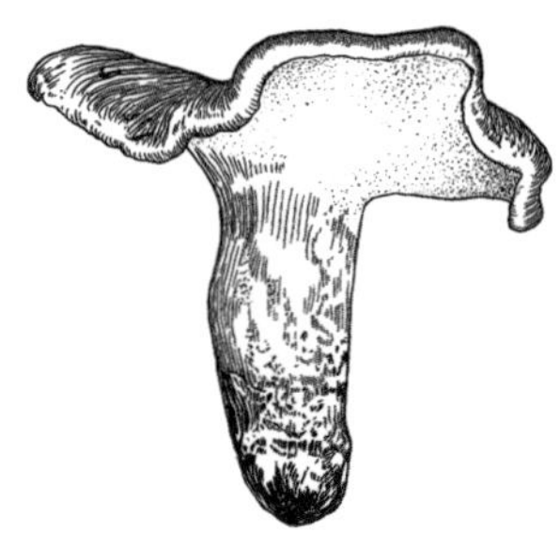

Boletopsis grisea

Boletus rex-veris

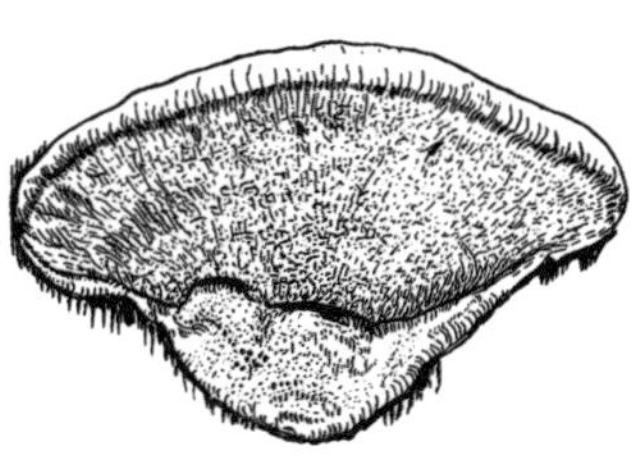

Hapalopilus nidulans

Hydnellum caeruleum

Hydnellum fuscoindicum

Hydnellum regium

Hydnellum suaveolens

Omphalotus olivascens

Phaeolus schweinitzii

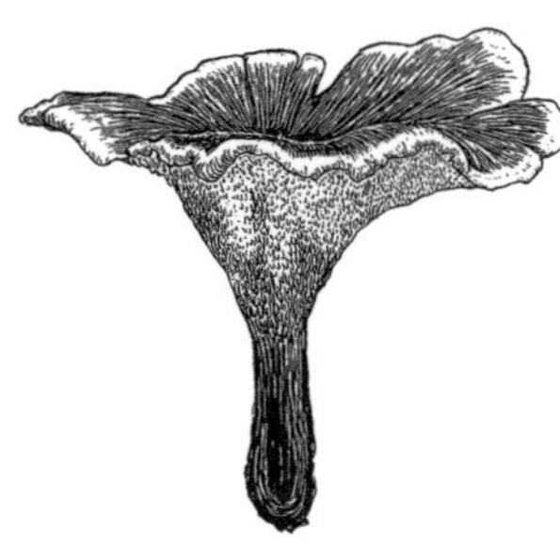

Phellodon niger

Polyozellus atrolazulinus

Sarcodon squamosus

Tapinella atrotomentosa

Xerocomellus zelleri

1. *Xerocomellus zelleri*, Pigment, Copper Acetate
2. *Omphalotus olivascens*, Pigment, Copper Acetate
3. *Aureoboletus mirabilis*, Pigment, Copper Acetate (Fresh)
4. *Boletus rex-veris*, Wool, Iron, pH5
5. *Sarcodon squamosus*, Pigment, Copper Acetate
6. *Tapinella atrotomentosa*, Pigment, Copper Acetate
7. *Hydnellum fuscoindicum*, Pigment, Soda Ash
8. *Boletopsis grisea*, Wool, Alum, pH9
9. *Hydnellum suaveolens*, Wool, Tin, pH9

10. *Hydnellum caeruleum*, Wool, Tin, pH9
11. *Polyozellus atrolazulinus*, Linen, Iron
12. *Hydnellum caeruleum*, Silk, Iron, pH9
13. *Polyozellus atrolazulinus*, Wool, Alum
14. *Polyozellus atrolazulinus*, Silk, Tin
15. *Polyozellus atrolazulinus*, Silk, Iron
16. *Hydnellum regium*, Linen, Iron, pH9
17. *Hydnellum regium*, Wool, Iron, pH9
18. *Hydnellum regium*, Silk, Iron, pH9

19. *Hydnellum fuscoindicum*, Wool, Iron, pH9
20. *Hydnellum fuscoindicum*, Linen, Iron, pH9
21. *Polyozellus atrolazulinus*, Linen, Titanium Oxalate
22. *Hydnellum regium*, Linen, Titanium Oxalate, pH9
23. *Hydnellum fuscoindicum*, Wool, Tin, pH9
24. *Hydnellum regium*, Silk, Alum, pH9
25. *Phellodon niger*, Pigment, Copper Acetate
26. *Hydnellum regium*, Pigment, Copper Acetate
27. *Hydnellum regium*, Wool, Alum, pH9

28 29 30

31 32 33

34 35 36

28. *Hydnellum fuscoindicum*, Pigment, Citric Acid
29. *Hydnellum fuscoindicum*, Pigment, Alum
30. *Hydnellum fuscoindicum*, Pigment, Base
31. *Boletopsis grisea*, Linen, Aluminum Acetate, pH9
32. *Hydnellum suaveolens*, Linen, Iron, pH9
33. *Boletopsis grisea*, Silk, Tin, pH9
34. *Hydnellum suaveolens*, Silk, Iron, pH9
35. *Hydnellum suaveolens*, Wool, Iron, pH9
36. *Hydnellum regium*, Linen, Aluminum Acetate, pH9

37. *Hydnellum fuscoindicum*, Wool, Alum, pH9
38. *Sarcodon squamosus*, Wool, Iron, pH9.5
39. *Tapinella atrotomentosa*, Wool, Iron (Fresh)
40. *Hydnellum suaveolens*, Pigment, Copper Acetate
41. *Tapinella atrotomentosa*, Silk, Iron (Fresh)
42. *Omphalotus olivascens*, Silk, Iron, pH5
43. *Boletopsis grisea*, Silk, Iron, pH9
44. *Boletopsis grisea*, Linen, Iron, pH9
45. *Sarcodon squamosus*, Silk, Iron, pH9.5

46. *Hydnellum caeruleum*, Pigment, Alum
47. *Hapalopilus nidulans*, Pigment, Copper Acetate
48. *Hydnellum fuscoindicum*, Silk, Iron, pH9
49. *Tapinella atrotomentosa*, Wool, Tin
50. *Phellodon niger*, Pigment, Soda Ash
51. *Boletopsis grisea*, Wool, Iron, pH9
52. *Phellodon niger*, Pigment, Base
53. *Phellodon niger*, Pigment, Alum
54. *Polyozellus atrolazulinus*, Wool, Tin

55. *Boletopsis grisea*, Wool, Tin, pH9
56. *Polyozellus atrolazulinus*, Pigment, Iron
57. *Hydnellum regium*, Wool, Tin, pH9
58. *Hydnellum caeruleum*, Pigment, Soda Ash
59. *Tapinella atrotomentosa*, Wool, Iron
60. *Omphalotus olivascens*, Wool, Iron, pH5
61. *Hydnellum caeruleum*, Pigment, Base
62. *Phaeolus schweinitzii*, Wool, Iron, pH9
63. *Phaeolus schweinitzii*, Wool, Iron, pH5

64. *Phaeolus schweinitzii*, Silk, Iron, pH5
65. *Hydnellum regium*, Pigment, Soda Ash
66. *Hydnellum regium*, Pigment, Iron
67. *Boletopsis grisea*, Pigment, Copper Acetate
68. *Boletopsis grisea*, Pigment, Iron
69. *Tapinella atrotomentosa*, Silk, Iron

CHAPTER FIVE

Blue

Indigo. Cobalt. Azure. Aqua. Sapphire. Ultramarine. Turquoise.

In studies conducted around the world, blue ranks as most people's favorite color. Perhaps it's the calming blue of the ocean reminding us we are mostly water, or the vastness of the blue sky above that draws us in. Blue is a primary color situated on the light spectrum between violet and cyan. It possesses a shorter wavelength—visible at 450 nanometers—than its primary color companions red and yellow. Blue's truncated wavelength scatters easily when colliding with particles in the atmosphere. This phenomenon, referred to as Rayleigh scattering, is the reason our sky appears blue. *The Blue Marble*, one of the most reproduced images in history, depicts our gleaming planet from 18,000 miles away, suspended in space. Captured in 1972 by the crew of *Apollo 17* while in transit to the moon, this iconic image helped to usher in a new era of environmental activism by depicting Earth as our blue planet.

The word *blue* is derived from the Middle English and Old French *bleu*, which can be traced back to the German word *blao*, meaning "shimmering" or "lustrous." Blue minerals, animals, and plants are rare. Blue flowers are produced by only 10% of flowering plant species. Only one known animal is thought to contain natural blue pigment—a species of butterfly, *Nessaea obrinus*. And there are few blue vegetables—even blue corn and blueberries are purple! Blue gemstones like sapphire are some of the rarest in the world, and vivianite ochre only develops its striking blue after undergoing oxidation.

Indigo is one of the most well-known and revered natural blue colorants. Its name refers both to the color and the plant. Indigo dyeing is an ancient art. Four thousand years ago in ancient Egypt, they threaded indigo-dyed fiber into the edges of linen burial cloths. In the late 1990s, textile historian Dr. Gillian Vogelsang-Eastwood identified a state robe worn by Tutankhamen as made almost entirely of indigo-dyed fiber. The historical significance of indigo touches many arenas—culture and the arts, trade and commerce, colonialism and slavery. In the 1740s, indigo was a major cash crop, second only to rice, in what is now South Carolina, and for the following fifty years it was grown on plantations where enslaved people were charged with the laborious process of growing, harvesting, and processing the "blue gold."[1] More recently, natural indigo has influenced environmental awareness and sustainable practices in the textile dyeing and denim industries. Artisans, farmers, and conscientious companies are using 100% plant-based indigo dye as a replacement for synthetic petrochemical-derived colorants.

The blues we come across most frequently are brilliant, intense, electric hues that are primarily derived from synthetic colorants. It is challenging to obtain blue colorants from nature, but the fungi kingdom is unique in that it offers a rare blue-producing pigment found nowhere else in nature.

---

1. Latria Graham, "The Blue that Enchanted the World," *Smithsonian*, November/December 2022, https://www.smithsonianmag.com/arts-culture/indigo-making-comeback-south-carolina-180980987.

These blues are subtle, earthy, and somewhat muted and irregular. They have hints of green and traces of gray, the result of multiple chemical pigment compounds combining within the mushrooms. These special compounds are known as terphenylquinones and occur in tooth fungi, polypores, and even some boletes. Achieving blue from mushrooms is not as simple as achieving other colors. The multiple pigment compounds that contribute to the creation of blue can also hinder its realization. Upon warming up the mushroom dye bath, immediately adjusting it to an alkaline (pH9) solution will help coax out the blue pigment. Unfortunately, blue colorants are fragile, and with only subtle changes to pH and oxygenation, they can be transformed into yellow-oranges or brown colors.

Each mushroom species responds differently to each variety of fiber and to the conditions in the dye pot. *Hydnellum caeruleum* creates a sky blue on linen and silk and teal blue on wool, whereas *Phellodon niger* makes an intense dark blue on silk and deep blue-green on wool. *Hydnellum fuscoindicum* produces a range of dazzling blues across various fiber types and mordants. The striking shades of blue-green from *Polyozellus atrolazulinus* (commonly known as the blue chanterelle) are as fabulous as the glittering blue-gray mushroom itself. A range of lovely blue dyes can be created from mushrooms, but blue pigments tend to shift greenish during the lake-making process, so the quest continues for a true blue pigment for paints and inks.

MUSHROOMS FEATURED IN THIS SECTION:

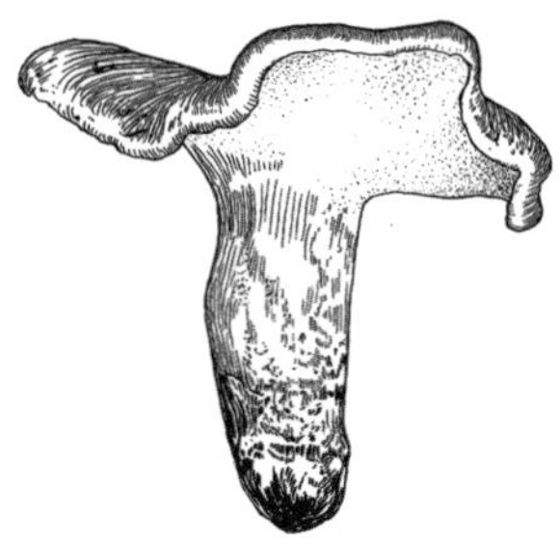

Boletopsis grisea

Boletus rex-veris

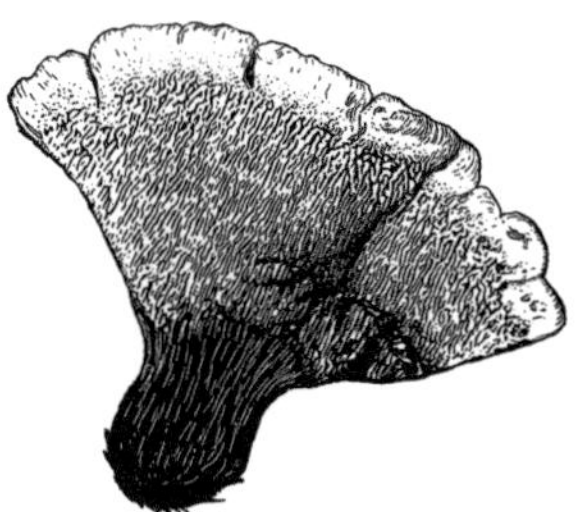

Hydnellum caeruleum

Hydnellum fuscoindicum

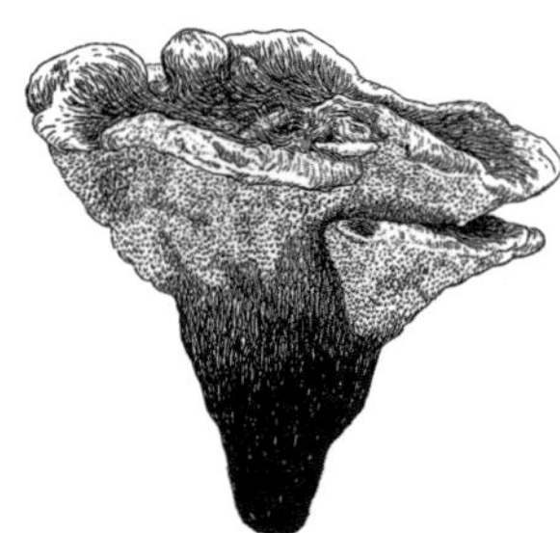

Hydnellum regium

Hydnellum suaveolens

Inonotus obliquus

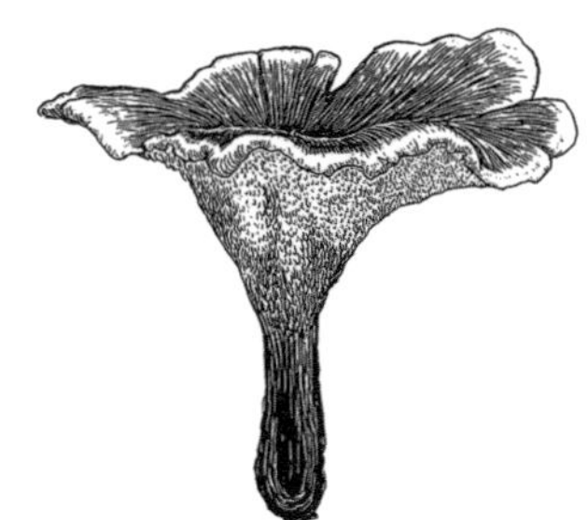

Phellodon niger

Polyozellus atrolazulinus

1. *Hydnellum suaveolens*, Linen, Titanium Oxalate, pH9
2. *Hydnellum suaveolens*, Silk, Tin, pH9
3. *Hydnellum suaveolens*, Linen, Aluminum Acetate, pH9
4. *Hydnellum fuscoindicum*, Pigment, Copper Acetate
5. *Hydnellum suaveolens*, Wool, Alum, pH9
6. *Hydnellum fuscoindicum*, Linen, Aluminum Acetate, pH9
7. *Hydnellum fuscoindicum*, Linen, Titanium Oxalate, pH9
8. *Hydnellum caeruleum*, Wool, Alum, pH9
9. *Hydnellum caeruleum*, Silk, Alum, pH9

10. *Hydnellum caeruleum*, Silk, Tin, pH9
11. *Hydnellum caeruleum*, Linen, Titanium Oxalate, pH9
12. *Hydnellum caeruleum*, Linen, Aluminum Acetate, pH9
13. *Phellodon niger*, Silk, Alum, pH9
14. *Phellodon niger*, Linen, Iron, pH9
15. *Phellodon niger*, Wool, Alum, pH9
16. *Phellodon niger*, Silk, Tin, pH9
17. *Phellodon niger*, Silk, Iron, pH9
18. *Phellodon niger*, Wool, Iron, pH9

19. *Hydnellum caeruleum*, Linen, Iron, pH9
20. *Polyozellus atrolazulinus*, Linen, Aluminum Acetate
21. *Polyozellus atrolazulinus*, Silk, Alum
22. *Hydnellum regium*, Silk, Tin, pH9
23. *Hydnellum caeruleum*, Wool, Iron, pH9
24. *Phellodon niger*, Linen, Titanium Oxalate, pH9
25. *Phellodon niger*, Linen, Aluminum Acetate, pH9
26. *Phellodon niger*, Wool, Tin, pH9
27. *Polyozellus atrolazulinus*, Pigment, Base

28. *Polyozellus atrolazulinus*, Pigment, Alum
29. *Polyozellus atrolazulinus*, Pigment, Copper Acetate
30. *Polyozellus atrolazulinus*, Wool, Iron
31. *Polyozellus atroalzulinus*, Pigment, Citric Acid
32. *Hydnellum regium*, Pigment, Alum
33. *Polyozellus atrolazulinus*, Pigment, Soda Ash
34. *Inonotus obliquus*, Pigment, Copper Acetate
35. *Boletus rex-veris*, Pigment, Copper Acetate (Fresh)
36. *Boletopsis grisea*, Linen, Titanium Oxalate, pH9

CHAPTER SIX

Purple

Violet. Lavender. Mauve. Plum. Lilac. Amethyst. Periwinkle. Eggplant.

Combining red and blue creates purple, the color with the shortest wavelength of any visible color at around 380 nanometers. Despite its characterization as a secondary color (sitting directly between primaries red and blue on the color wheel), purple has historically been a signifier of royalty, divinity, and decadence. There are records of Roman and Persian emperors wearing purple tunics more than 2,000 years ago. In Elizabethan England, only members of the royal family were permitted to wear purple garb. Purple dyes were symbols of status because they were so rare and difficult to create—in ancient Rome, this coveted color was accessible only to the elite.

As early as the twelfth century BCE, Tyrian purple was derived from the mucus glands of a predatory sea snail (*Bolinus brandaris*) harvested off the coast of present-day Lebanon. Producing Tyrian purple was a tedious process—each snail was "milked" to produce a single drop of pigment. An ounce of this dye required the processing of roughly 250,000 snails!

The etymology of the word *purple* can be traced back to this natural dye: *Porphyra*, the ancient Greek word for purple, is derived from an older Eastern Mediterranean word for the sea snail responsible for Tyrian purple. Another sought-after purple pigment was derived from a rock-clinging lichen, *Roccella tinctoria* (commonly known as orchil), which also required a laborious and intensive extraction process.

Purple is also a rare colorant within the fungi kingdom. There are only a few mushrooms that can create purples, and the pigments responsible for these shades are unique to mushrooms.[1] Whether they are made from the pigments alone or in conjunction with tannins and irons, mushrooms' purple values range from lovely, bright lilac to smoky, earthy violet-gray. Elsewhere in the natural world, purple is most widely expressed by flowers such as lilacs, lavender, and wisteria. Purple fruits and vegetables such as cabbage, eggplant, and plums receive their color from anthocyanins, a compound that may also contribute health benefits.

British chemist William Henry Perkin accidentally discovered the color mauve in his lab in 1856, forever changing the history of colorants. Perkin's mauve was the first dye created from coal tar, and it revolutionized the advancement of synthetic dyes, leading to the demise of using natural dyes at scale. These bright, garish synthetic colors exude artificiality, especially when compared to the complex purple hues derived from natural sources like mollusks, lichens, and mushrooms.

The purple terphenylquinone pigments found in the fungi kingdom include polyporic acid, thelephoric acid, and atromentin, all of which create a variety and range of purples and violets. These colors can prove elusive. Some mushrooms require an alkaline (pH9) shift while others require an acidic (pH5) shift to coax out compelling

---

1. Räisänen, Primetta, Niinimäki, *Dyes from Nature*, 199.

shades. The combination of fiber choice and mordant will also impact whether you manage to create a purple or green color. Purple hues are often improved with mordants such as alum, iron, and tin. *Tapinella atrotomentosa*—classified as a bolete even though it has gills under its cap—produces some beautiful shades of light, smoky lavender and has earned itself the nickname "temperamental *Tapinella*" due to the exacting conditions it requires to yield those marvelous shades.

One of the most coveted dye mushrooms is *Hapalopilus nidulans* (commonly known as cinnamon bracket), which consistently creates an Easter egg purple. This nesting polypore is prevalent throughout Scandinavia and can be found on the east coast of North America as well as in Africa and Asia. When foraging for mushrooms in the field, mycologists will use reagents such as KOH, which is a 3–10% solution of potassium hydroxide (don't get any on your skin or it will burn!) to further aid with identification. Putting a drop of KOH on *H. nidulans* will yield a purple stain on the mushroom, indicating it contains the polyporic acid pigment. The Western jack-o'-lantern mushroom, *Omphalotus olivascens*—found only in California and a small section of Northern Mexico—not only glows in the dark with its own bioluminescence but also produces a beautiful range of deep, rich purples. The genus *Ramaria*, the colorful coral mushroom, can create smoky violets in combination with iron mordant.

MUSHROOMS FEATURED IN THIS SECTION:

Cortinarius neosanguineus

Cortinarius ominosus

Cortinarius subcroceofolius

Cortinarius uliginosus

Echinodontium tinctorium

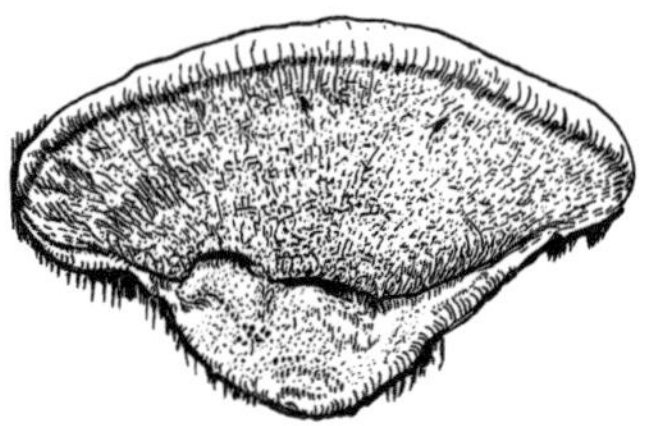

Hapalopilus nidulans

Hypomyces lactifluorum

Omphalotus olivascens

Ramaria spp.

Sarcodon squamosus

Tapinella atrotomentosa

1. *Hapalopilus nidulans*, Pigment, Soda Ash
2. *Hapalopilus nidulans*, Wool, Alum
3. *Hapalopilus nidulans*, Wool, Tin
4. *Hapalopilus nidulans*, Pigment, Citric Acid
5. *Hapalopilus nidulans*, Pigment, Base
6. *Hapalopilus nidulans*, Pigment, Alum
7. *Cortinarius neosanguineus*, Linen, Iron, pH10
8. *Cortinarius neosanguineus*, Wool, Iron, pH10
9. *Cortinarius neosanguineus*, Silk, Iron, pH10

10. *Omphalotus olivascens*, Wool, Alum, pH5
11. *Cortinarius ominosus*, Silk, Iron, pH9
12. *Ramaria* spp., Silk, Iron
13. *Sarcodon squamosus*, Silk, Alum, pH9.5
14. *Sarcodon squamosus*, Wool, Alum, pH9.5
15. *Sarcodon squamosus*, Wool, Tin, pH9.5
16. *Cortinarius neosanguineus*, Linen, Iron
17. *Cortinarius subcroceofolius*, Pigment, Copper Acetate
18. *Hypomyces lactifluorum*, Wool, Iron, pH10

19. *Cortinarius uliginosus*, Pigment, Copper Acetate
20. *Echinodontium tinctorium*, Pigment, Soda Ash
21. *Omphalotus olivascens*, Silk, Tin, pH5
22. *Cortinarius ominosus*, Linen, Iron, pH9
23. *Ramaria* spp., Wool, Iron
24. *Cortinarius ominosus*, Wool, Iron, pH9
25. *Ramaria* spp., Linen, Iron
26. *Hapalopilus nidulans*, Wool, Iron
27. *Omphalotus olivascens*, Silk, Alum, pH5

28. *Echinodontium tinctorium*, Wool, Iron, pH10
29. *Tapinella atrotomentosa*, Silk, Alum
30. *Tapinella atrotomentosa*, Wool, Alum
31. *Tapinella atrotomentosa*, Wool, Alum (Fresh)
32. *Tapinella atrotomentosa*, Silk, Tin
33. *Tapinella atrotomentosa*, Silk, Tin (Fresh)

SECTION II

MUSHROOMS

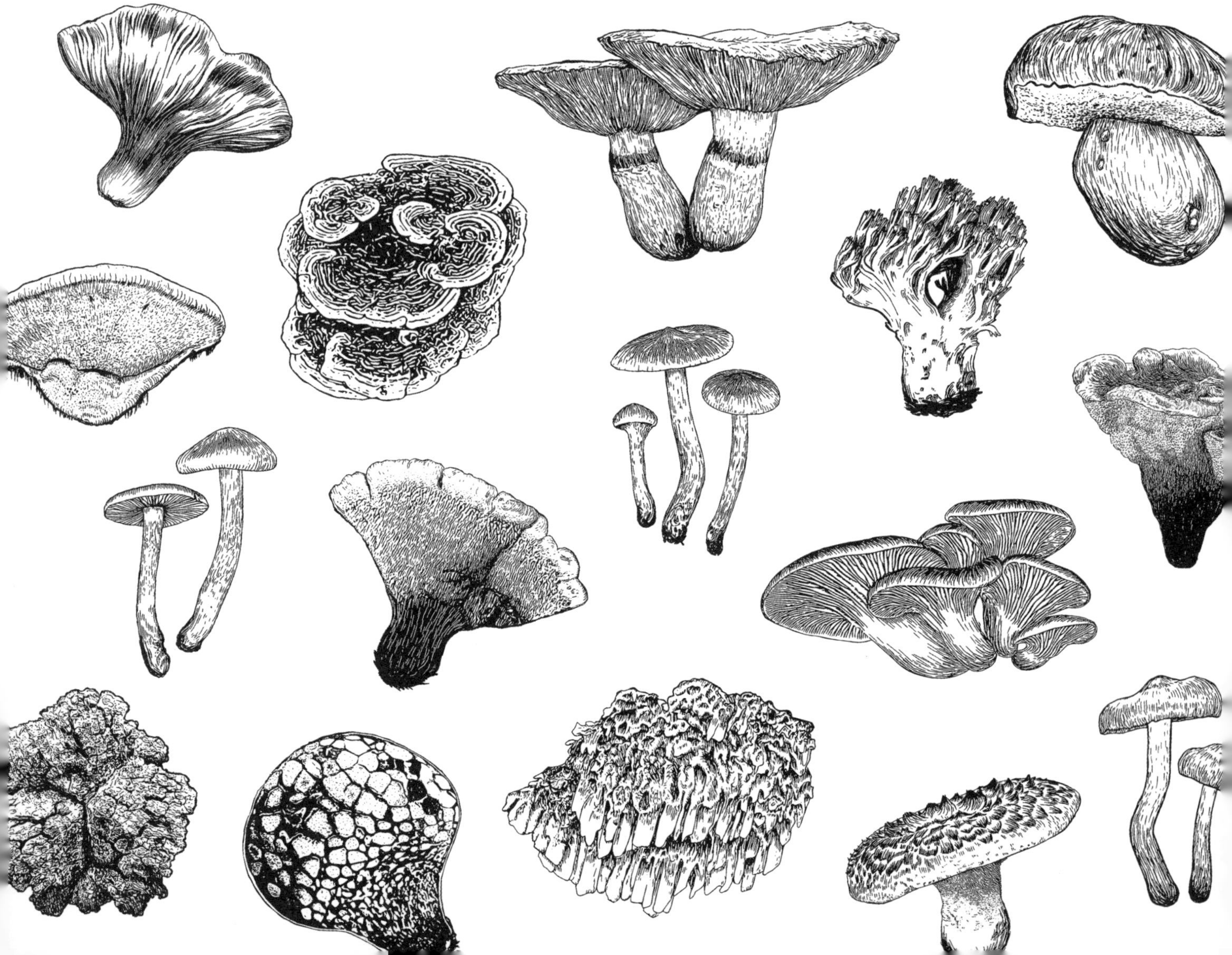

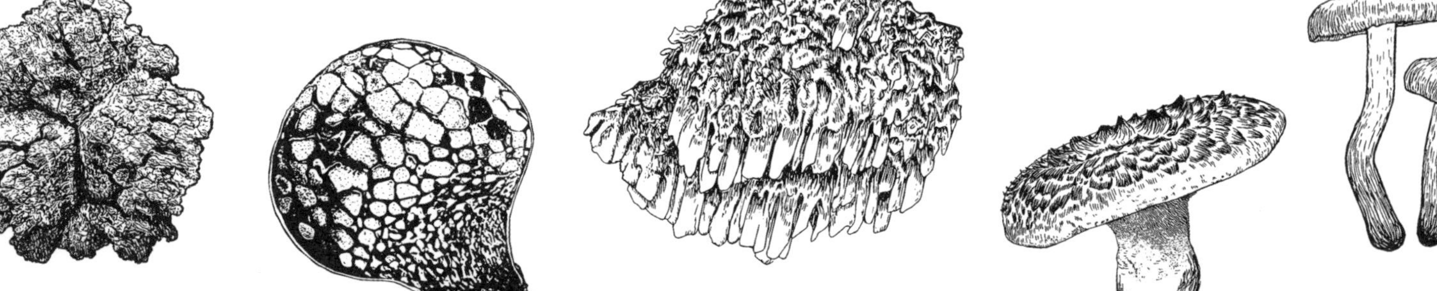

Fungi are an incredibly adaptable and widespread kingdom of life forms. They thrive on every continent, in every climate, from snowy peaks and sun-bleached deserts to the depths of the ocean floor. Mushrooms are neither plants nor animals but the fruits of underground thread-like fungal bodies called mycelium. As nature's primary mechanism for breaking down organic matter, fungi are critical to the health of our environment. They excel at returning nutrients to the soil and keep our planet from becoming overwhelmed by heaps of dead plants. In addition to being beautiful and delicious, medicinal and poisonous, humble and otherworldly, mushrooms are a great source of pigments that can be transformed into dyes, inks, and paints.

The fungi kingdom comprises a wild menagerie of organisms from yeasts and molds to lichens and mushrooms. This book specifically showcases mushrooms as a source of pigment and colorant (although colors can also be derived from lichens, bacteria, and fungal microorganisms, some of which can even be grown at scale to produce colorants for various industries). The mushrooms (with the exception of one, which is not technically a mushroom!) featured in this book are classified within the phylum Basidiomycota, meaning they exhibit the stereotypical "mushroomy" shape: stipe, cap, and spore-bearing surface (hymenium). In the world of biological taxonomy, each species is given a two-part Latin name, where the first part of the name denotes the genus and the second part of the name is the species within the genus.

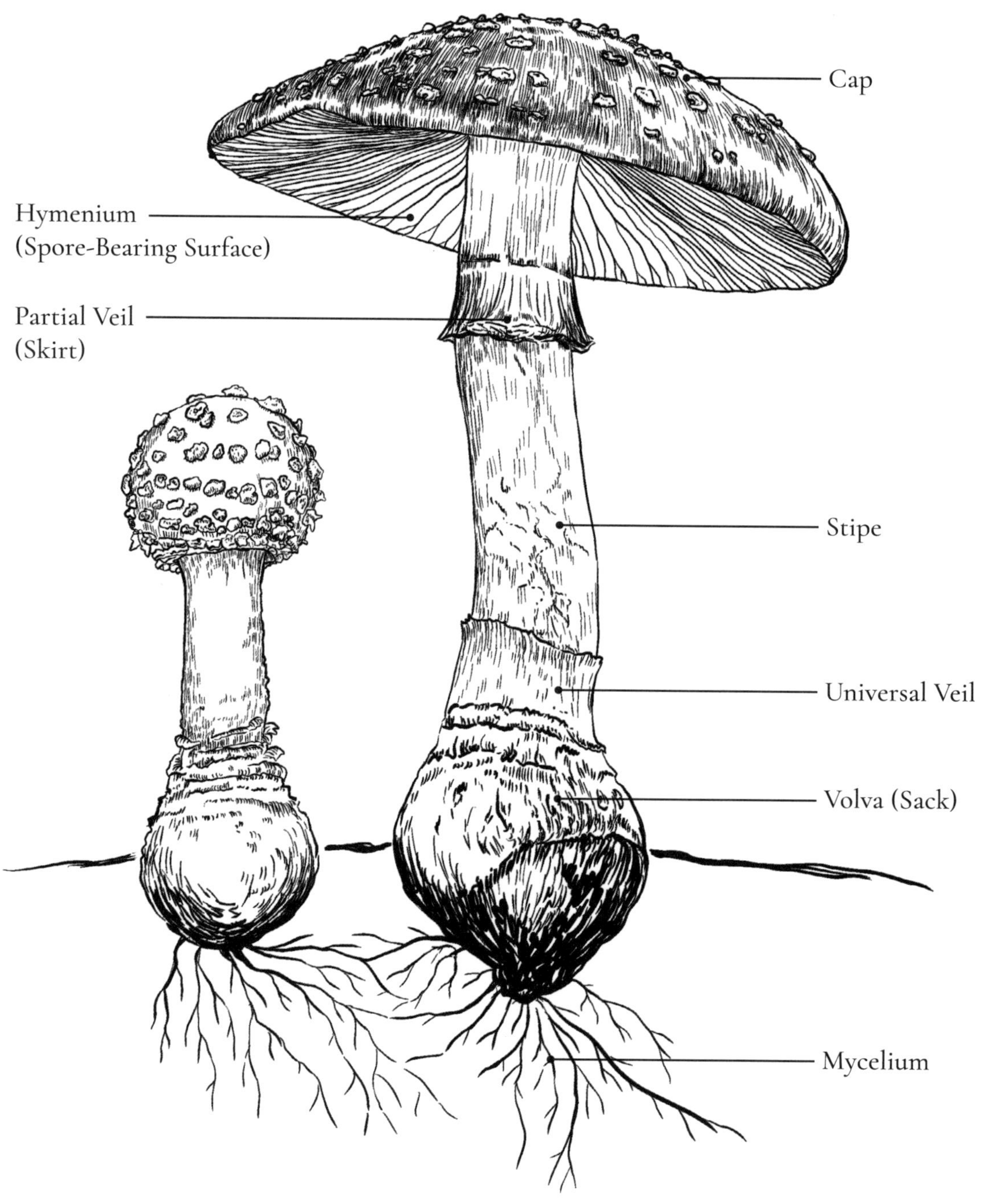
Cap
Hymenium
(Spore-Bearing Surface)
Partial Veil
(Skirt)
Stipe
Universal Veil
Volva (Sack)
Mycelium

You may be able to find wild mushrooms that are both edible and good for dyeing at local farmers' markets, and it is even possible to cultivate some species of mushrooms that produce colorants in your own home. Growing your own dye mushrooms would be a satisfying and elaborate process that is beyond the scope of this book, but I encourage you to explore fungi cultivation if you have the interest. However, for me, one of the more satisfying parts of producing colorants from mushrooms is foraging for pigment-bearing mushrooms in the wild. Learning to identify mushroom species with the help of trusted experts from local mycological societies and published field guides is an empowering process to becoming a conscientious and sustainable forager. Where necessary, obtain the proper permits required to collect wild mushrooms. Be careful about trespassing, and only forage in approved areas. As with any type of wild foraging, be judicious and ensure your specimens are not on the International Union for Conservation of Nature's (IUCN) Red List of Threatened Species. Since mushrooms are the fruiting bodies of fungi, like the fruit of an apple tree, harvesting mushrooms in quantity does not injure the main fungal organism as long as you are careful not to damage the mycelium beneath—doing so would be the equivalent of cutting a branch off an apple tree just to harvest a few apples. To help keep the mycelium intact, one approach is not to pull large mushrooms out of the ground. Instead, use a small sharp knife to gently cut through a mushroom's stalk, known as its stipe, near its base.

Ethical foragers only take what they need. We know that mushrooms improve the overall biodiversity of their environment, so it's important to leave some mushrooms for the animals and insects that depend on them for food (plus, it's kind to leave some for other human foragers as well). Harvesting only mature mushrooms that have already had the opportunity to disperse their spores helps maintain the long-term abundance of the species. Likewise, leaving a few fruiting bodies behind helps ensure spores will produce new mycelium for the next season. If you are unable to identify a mushroom in the field, take home only a few samples to aid in your identification process. I think you will be surprised how few mushrooms you need to make colors. The maturity of the mushrooms you collect will significantly impact the dye: Young specimens of some species will provide pigment, but it's typically the older, fully mature mushrooms that produce the best colorants.

Learning the visual characteristics of different mushroom species is critical for proper identification. However, I find getting close to a mushroom and opening my senses to explore the minute details of its form, its texture, and even its smell can help make identification easier. And don't worry: It is safe to touch any mushroom, even if it is toxic! Just be sure not to eat any mushroom unless you are 100% confident in your identification! Some people like to make spore prints on paper, with the color of the shed spores providing another data point toward identification. Sometimes, even with all my research and observation skills, I'm still not certain about a mushroom's identity, and further study with a microscope or DNA sequencing is necessary. Understanding the habitats where each species tends to fruit is one of the most important clues to identification: What types of trees does it grow under? Does it prefer an open field? Does it fruit from living or dead wood, on branches or trunks? Or does it grow directly out of the soil? Does it grow solitary, scattered, or in clusters?

Observing details about a mushroom's cap, stipe, and hymenium helps ensure a positive identification. Almost every mushroom in this book has some form of stipe—also referred to as a stalk or stem—connecting its cap to its subterranean mycelium. Stipes range in size and shape as well as in the way they attach to the cap or side of a tree. Learning the nuances of each species' stipe will help you home in on a positive ID. Likewise, the color, size, shape, and texture of a mushroom's cap offers significant clues. However, it is the underside of the cap—specifically the hymenium—that is often the most important identifying feature of any mushroom. This spore-bearing surface is one of the main criteria used to place mushrooms into taxonomic groups.

The Mushroom chapters in this book are organized by type: Boletes, Gilled, Polypores, Tooth, and Odds & Ends. Within each chapter, illustrations show the mushroom at different stages of life and from multiple views. The illustrations are accompanied by detailed identification information for each species. Each mushroom also has its own record page, which displays the spectrum of hues extracted from that particular mushroom, along with the simple formula used to achieve the dyes and pigments. Each formula considers the following variables:

- Fiber pre-treatment and mordants
- Condition of the mushroom
- Mushroom parts used
- Ratio of weight of goods to weight of fiber (WOG:WOF)
- Length of time cooking
- Temperature of dye bath
- pH of dye bath

CHAPTER ONE

Boletes

I will never forget the first time I spied a porcini mushroom—*Boletus edulis*—fruiting proudly at the base of a spruce tree. I'd been tromping through the national forest close to my home for a few hours hunting for fungi. I froze when I spotted a stout, tawny-capped specimen among the spruce needles. I instantly recognized it from poring over mushroom guides—I had dreamed of this moment. I approached slowly with wide eyes. I squatted down and touched the mushroom's smooth, soft cap and ran my fingers along its pored underside. I admired the gorgeous reticulating netting pattern decorating its strong stipe. I took out my mushroom knife and gently harvested it, careful not to disturb the mycelium, the living network from which it had sprung, at its base. I held this porcini close to my face and smelled its aroma of sweet soil. When I cut the mushroom open, its interior flesh was clean, crisp, and white. Once I had checked all the identification boxes, I let out a little shriek of delight. *My first porcini!*

Boletes are a large and popular genus of mushrooms that includes some of the most delicious edible species. Members of the *Boletus* genus tend to be thick, sturdy looking, and come in many colors: often tan, brown, yellow, orange, or red. Some species quickly bruise blue or green when touched. On the undersides of their fleshy caps, where many mushrooms have gills, boletes have a sponge layer composed of a multitude of pores that can appear smooth when young. These pores are tightly packed tubes that hold and release the mushroom's reproductive spores. The tube pores can be easily separated from the cap, a characteristic that distinguishes boletes from other polypores. While the name *bolete* is often used generically to describe various mushrooms, this practice can lead to confusion for novice foragers. It is worthwhile to learn about the unique species that grow in your area.

The hues produced from boletes range from lemony yellow to gold to mustard orange to olive green. The spore tubes themselves contain a high concentration of pigment and bear most of the color-producing chemicals in the *Boletus* genus. To coax more brilliant colors from these mushrooms, simply add household distilled white vinegar to the dye bath to make it more acidic (pH4). Selecting older, more mature mushroom specimens is the key to dye potency. Boletes often make a murky mess of the dye pot, and I find I must strain out the last little bits to ensure clean pigment and fiber. Working with dried bolete mushrooms might prove easier, but fresh or frozen specimens often create more vibrant colors.

Bolete species are found all over the world in a range of climates and biomes. In fact, the genus is so large and expansive that in recent years it has been further subdivided into separate genera. For example, *Aureoboletus* and *Xerocomellus* are each an example of a bolete mushroom that has been placed in its own genus. *Aureoboletus* has bright yellow pores with a tapered stipe base. *Xerocomellus* has yellow pores, reddish-yellow stipes, and often stains blue. *Boletus rex-veris* is a well-known edible mushroom commonly referred to as the spring king. This close cousin of the delicious porcini (*Boletus edulis*) also has great dyeing potential, especially when it reaches a mature age. Curiously, the species *Tapinella atrotomentosa* is a gilled bolete! This mushroom doesn't have the typical sponge layer of tubelike pores, but if you were to study *Tapinella*'s individual spores under a microscope, you'd find its elongated spore shape ties it to the rest of the *Boletus* genus.

MUSHROOMS FEATURED IN THIS SECTION:

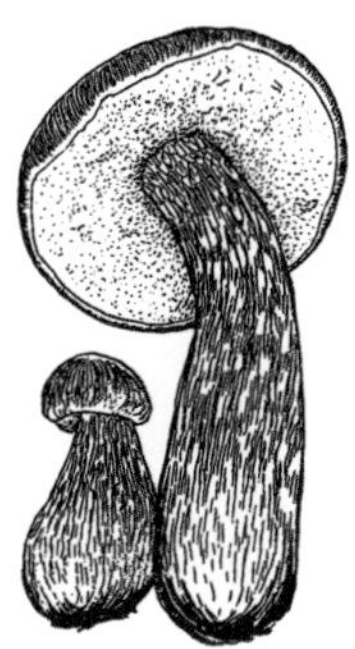

Aureoboletus mirabilis

Boletus rex-veris

Tapinella atrotomentosa

Xerocomellus zelleri

Aureoboletus mirabilis

GENUS: *Aureoboletus*

SPECIES: *mirabilis*

COMMON NAME: Admirable Bolete

Aureoboletus mirabilis has a dark red-brown plush cap with yellowish pores and a dark red-brown stipe that is long and clublike in shape. Most of the pigment is concentrated in the cap; however, the flesh of this mushroom can be marbled red and yellow. All parts of the mushroom can be used to obtain color.

HABITAT & ECOLOGY: Solitary or scattered, growing on well-decayed, mossy wood. Primarily found with hemlocks.

DISTRIBUTION: Asia, West Coast of North America

SPORE PRINT: Dark olive-brown

CAP: Dark red to red-brown with a yellowish overhanging narrow band around the edge. Pinkish to white spots that appear as the mushroom matures.

HYMENIUM: Sponge layer has pores known as tubes, which are light yellow to bright yellow, becoming olive-yellow with age.

STIPE: Club-shaped with a large base. Dark brown to red-brown with marbled streaks of tan-pink.

Dye Preparation

CONDITION	Fresh
PART	Caps
PRE-SOAK	–
pH	5
RATIO	1:1
TEMP	160°F/71°C
TIME	1 hour

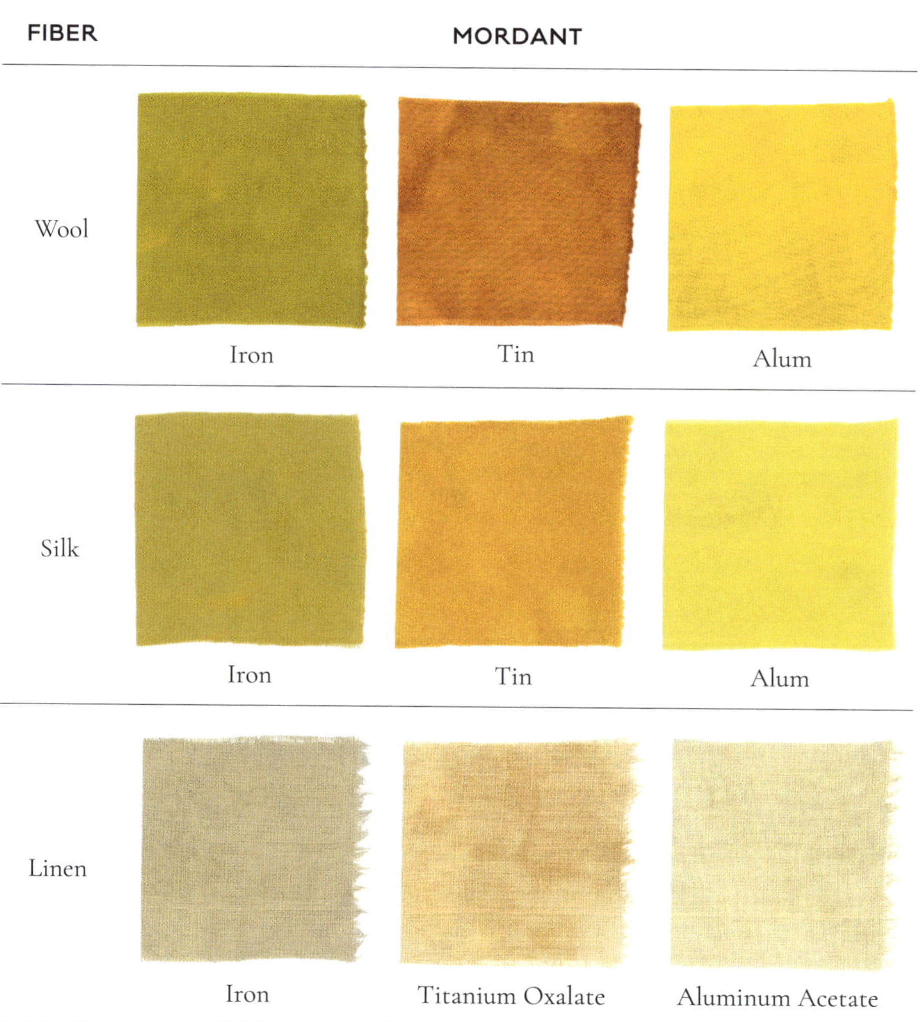

Pigment Preparation

CONDITION	Fresh
PART	Caps
RINSE	Twice
BINDER	Gum Arabic
TEMP	85°F/29°C

PIGMENT

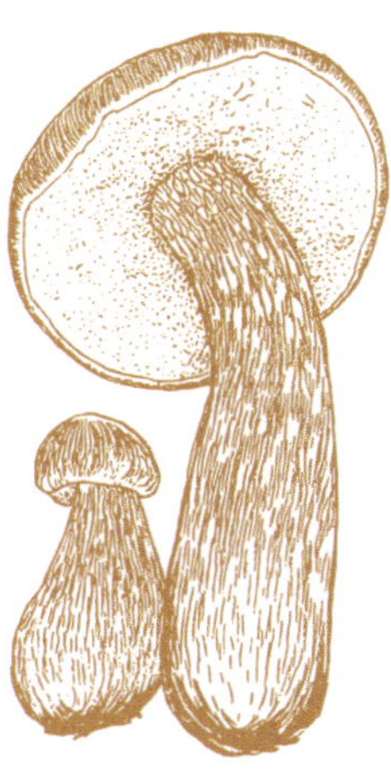

Dye Preparation

CONDITION	Dried
PART	Entire mushroom
PRE-SOAK	–
pH	5
RATIO	2:1
TEMP	165°F/74°C
TIME	1 hour

Pigment Preparation

CONDITION	Dried
PART	Entire mushroom
RINSE	Twice
BINDER	Gum Arabic
TEMP	85°F/29°C

PIGMENT

Base

Alum

Citric Acid

Iron

Copper Acetate

Soda Ash

Boletus rex-veris

GENUS: *Boletus*

SPECIES: *rex-veris*

COMMON NAME: Spring King

A kissing cousin of *Boletus edulis*, *Boletus rex-veris* has a tan-pink to light brown cap, and its stipe is often large and swollen with a netlike reticulation pattern, especially at the top. The pores under the cap are white, becoming yellow to olive-brown with age. The pigment is concentrated in the cap, specifically in the tubelike pores that can easily be peeled off from the cap.

HABITAT & ECOLOGY: Solitary or in clusters, growing under conifers, specifically pines and true firs. Often hidden under duff.

DISTRIBUTION: Western North America

SPORE PRINT: Olive-brown

CAP: Young unexposed caps are light tan, turning tan-pink then red-brown to dark red-brown with age.

HYMENIUM: Sponge layer has pores known as tubes, which are creamy white, becoming light yellow and eventually olive-yellow to olive-brown with age.

STIPE: Swollen stipe with an enlarged base that is cream colored, gradually becoming light brown with reticulation.

Dye Preparation

CONDITION	Fresh
PART	Caps
PRE-SOAK	–
pH	5
RATIO	2:1
TEMP	165°F/74°C
TIME	1 hour

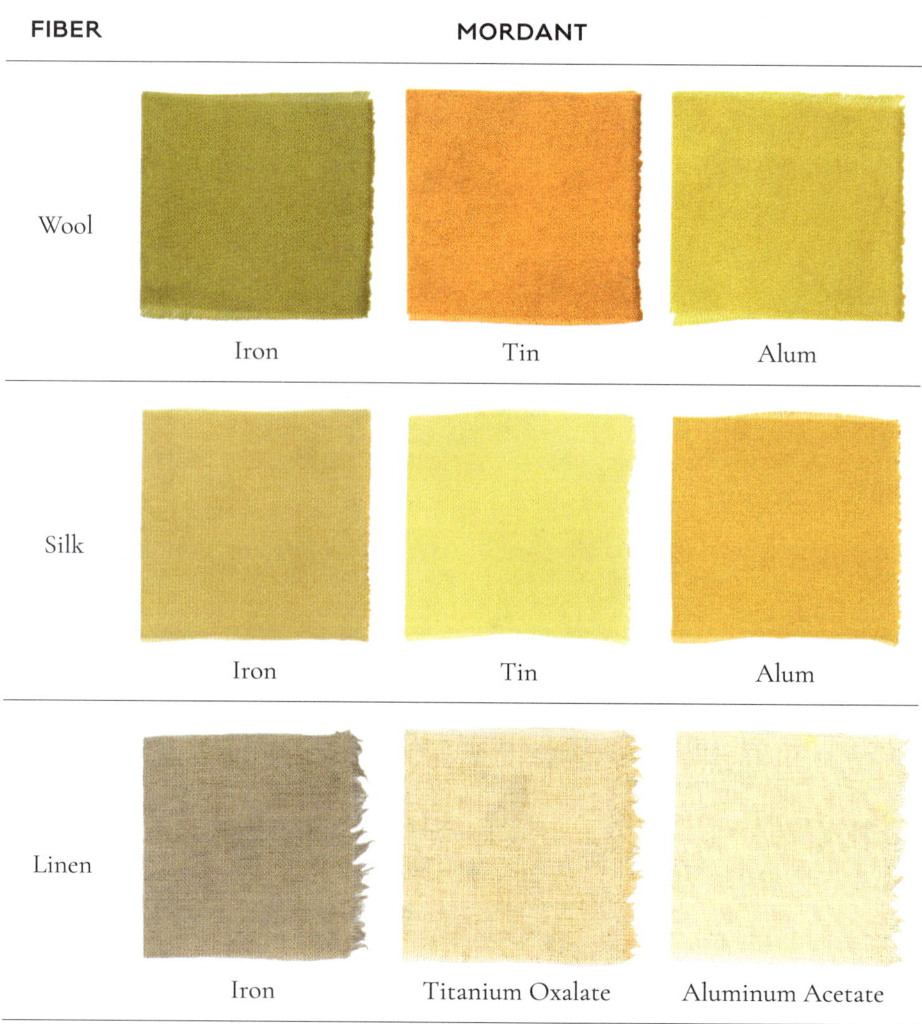

Pigment Preparation

CONDITION	Fresh
PART	Caps
RINSE	Twice
BINDER	Gum Arabic
TEMP	85°F/29°C

PIGMENT

Base

Alum

Citric Acid

Iron

Copper Acetate

Soda Ash

Dye Preparation

CONDITION	Dried
PART	Caps
PRE-SOAK	–
pH	5
RATIO	2:1
TEMP	165°F/74°C
TIME	1 hour

Pigment Preparation

CONDITION	Dried
PART	Caps
RINSE	Twice
BINDER	Gum Arabic
TEMP	85°F/29°C

PIGMENT

Base | Alum | Citric Acid

Iron | Copper Acetate | Soda Ash

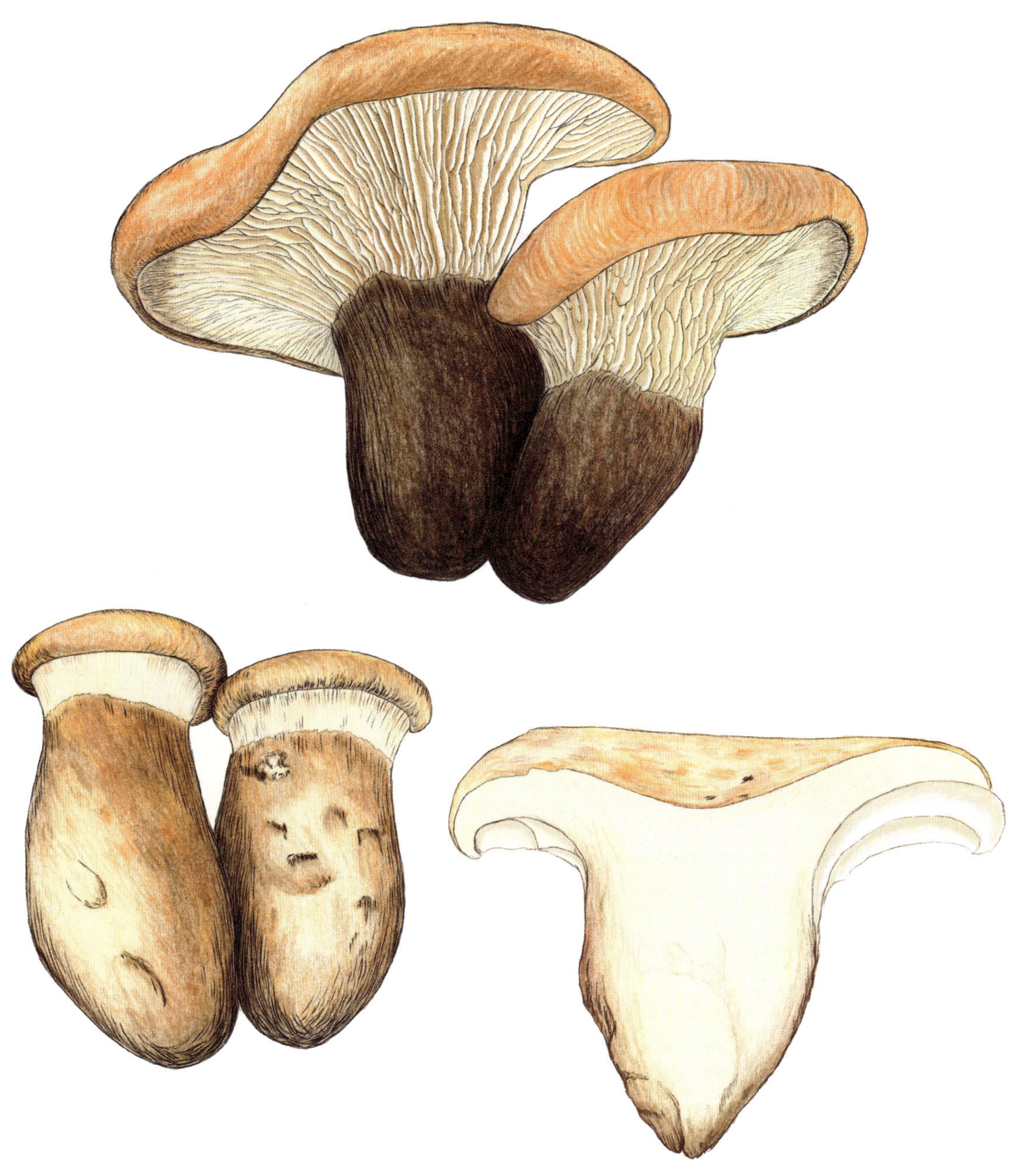

Tapinella atrotomentosa

GENUS: *Tapinella*

SPECIES: *atrotomentosa*

COMMON NAME: Velvet Rollrim

Tapinella atrotomentosa has a dry, suede-like light brown cap, and its stipe is thick with a velvety dark brown surface. It is known as "temperamental *Tapinella*" because the conditions have to be just right to create a beautiful purple.

HABITAT & ECOLOGY: Solitary, scattered, or in clusters growing on rotting conifer stumps, logs, or buried roots.

DISTRIBUTION: Asia, Central America, Europe, North America

SPORE PRINT: Yellow-brown

CAP: Light brown to dark brown, becoming uplifted and wavy with age. Suede-like velvet to touch and cracking with age.

HYMENIUM: The gills are creamy beige to yellow-brown, often forking and easy to scrape off.

STIPE: Thick, rather short, and off-center with a velvety dark brown to black surface with age.

Dye Preparation

CONDITION	Fresh
PART	Entire mushroom
PRE-SOAK	–
pH	7
RATIO	1:1
TEMP	100°F/38°C
TIME	1.5 hours

Pigment Preparation

CONDITION	Fresh
PART	Entire mushroom
RINSE	Twice
BINDER	Gum Arabic
TEMP	85°F/29°C

PIGMENT

Base

Alum

Citric Acid

Iron

Copper Acetate

Soda Ash

Dye Preparation

CONDITION	Dried
PART	Entire mushroom
PRE-SOAK	–
pH	7
RATIO	1:1
TEMP	100°F/38°C
TIME	1 hour

Pigment Preparation

CONDITION	Dried
PART	Entire mushroom
RINSE	Twice
BINDER	Gum Arabic
TEMP	85°F/29°C

PIGMENT

Base

Alum

Citric Acid

Iron

Copper Acetate

Soda Ash

Xerocomellus zelleri

GENUS: *Xerocomellus*

SPECIES: *zelleri*

COMMON NAME:
Zeller's Bolete

Xerocomellus zelleri has a dark purple to dark red dry, wrinkled cap with a light, pale yellow band around the edge, while the very similar *Xerocomellus atropurpureus* tends to lack the pale band and has a thicker stipe. Creamy yellow pores are visible under the cap, and the stipe is rosy red to dark red. Most of the pigment is concentrated in the cap (either species); however, the flesh of this mushroom is marbled yellow. All parts can be used to obtain color.

HABITAT & ECOLOGY: Solitary or scattered in moss or around moss-covered stumps and logs. Growing with a wide variety of trees, from conifers to hardwoods.

DISTRIBUTION: Western North America

SPORE PRINT: Olive-brown

CAP: Dark purple to dark red with a smooth, velvety, dry, wrinkled surface.

HYMENIUM: Sponge layer has pores known as tubes, which are light yellow to bright yellow, becoming dingy olive-yellow with age.

STIPE: Cylindrical with an enlarged base. The yellowish base is covered with rosy red to dark red streaks. Yellow-red becomes dark red with age.

Dye Preparation

CONDITION	Dried
PART	Caps
PRE-SOAK	–
pH	5
RATIO	2:1
TEMP	160°F/71°C
TIME	1 hour

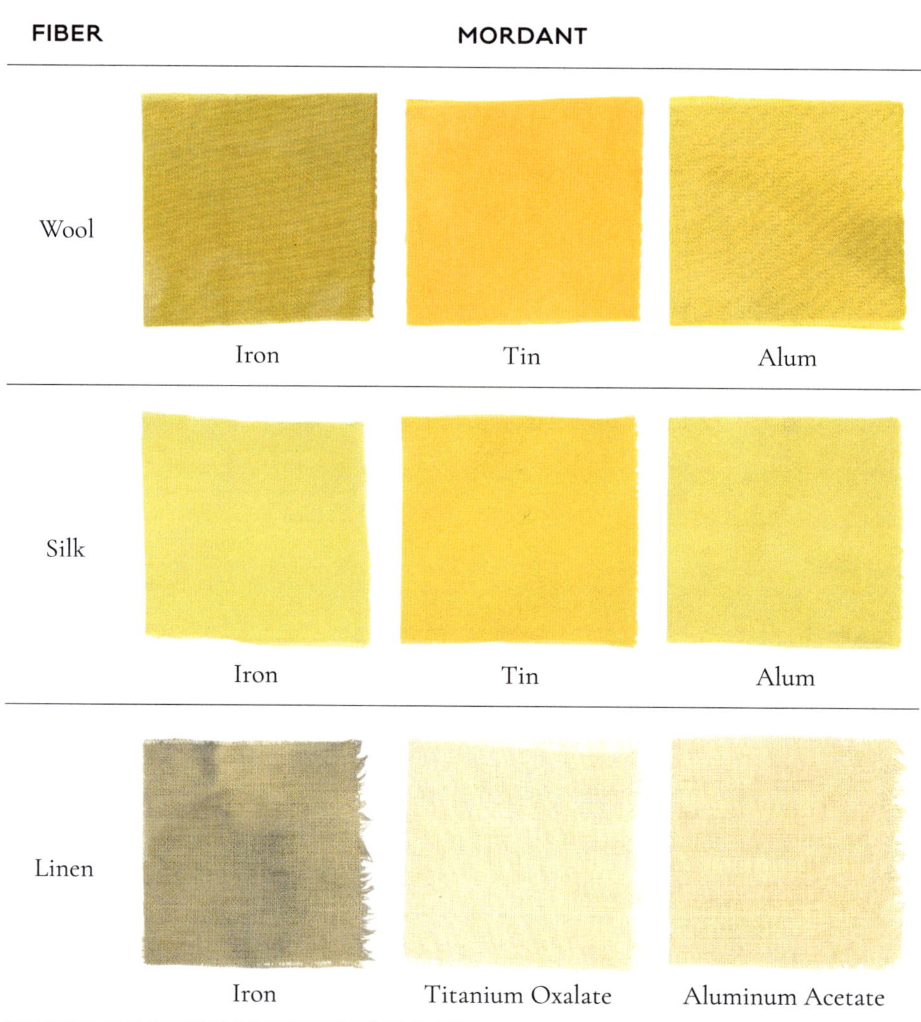

Pigment Preparation

CONDITION	Dried
PART	Caps
RINSE	Twice
BINDER	Gum Arabic
TEMP	85°F/29°C

PIGMENT

Base | Alum | Citric Acid

Iron | Copper Acetate | Soda Ash

CHAPTER TWO

Gilled

Most gilled mushrooms have the traditional mushroom shape with their stipe, cap, and gilled hymenium. The majority of mushrooms have gills under their caps, including many of the mushrooms we commonly eat, like shiitake, oyster, and portobello. Gilled mushrooms can be further classified by the color of their spore print: light-spored, pink-spored, brown-spored, or dark-spored. One of the most coveted mushrooms for pigment is the subgenus *Cortinarius dermocybe*, which primarily bears rusty brown spores.

When I started identifying mushrooms, I noticed the acronym LBM—"little brown mushroom"—was thrown around whenever someone didn't know the genus and species of a little brown mushroom (of which there are many!). The various dye mushrooms within the *C. dermocybe* subgenus all looked distinct and brightly colored in my field guides. But when I took my identification skills into the field, I struggled to differentiate between a *C. dermocybe* and any number of LBMs. Everything I thought was *C. dermocybe* turned out, after critical observation and identification, not to be! Despite my frustration, I persisted. Late one fall, when I was out with my mom foraging in habitats that I hoped would prove fruitful for these gems, I spotted a patch of a hundred or so *Cortinarius subcroceofolius*. My heart fluttered, but I knew better than to jump to conclusions, so I calmly walked up and evaluated a few of them. They looked promising. I harvested a couple, then sliced one open. I immediately recognized the telltale signs: a cobweb-like netting structure, a solid stipe (not hollow), and rust-brown spores. I let out a huge *hooray* that startled my mom. She ran over, and I showed her the identification characteristics of the species. A few hours later we found ourselves with two bags full of these fabulous mushrooms!

C. dermocybe contains some of the best pigment compounds: anthraquinones. The most famous ancient red dyes such as madder, cochineal, and kermes all contain anthraquinones. *C. dermocybe* bears its namesake pigment, dermocybin, which creates a brilliant red with *Cortinarius neosanguineus* (using the stipe and caps) and *Cortinarius ominosus* (using caps only). These dyes are improved with mordants. Dermorubin is the light rose pigment found in *Cortinarius uliginosus*, and emodin—probably the most widely distributed anthraquinone—produces varying shades of orange in *C. subcroceofolius*. Shifting the pH of the dye bath can push the pigmented colors into different hues, including purple. An alkaline (pH9) dye bath pushes colors toward red and pink, whereas an acidic (pH5) environment pushes colors toward orange and yellow.

Many other gilled mushrooms besides *Cortinarius* contain good pigments. *Hypomyces lactifluorum*, commonly known as the lobster mushroom, is one of the most intriguing. The lobster mushroom is usually a classic *Russula brevipes* mushroom (and occasionally other species) that has a parasitic fungus, *Hypomyces lactifluorum*, living within its tissues. This parasitism turns the *Russula* a reddish-orange color that resembles the shell of a cooked lobster. Since it is this brightly colored surface layer that holds the pigment compounds, I will often peel it off and toss it into the dye pot, discarding the thick white inner flesh. These beautifully flask-shaped mushrooms are a choice edible for many people, but I much prefer the older, more mature specimens for the dye pot. They can create an array of hues when processed, fresh or dried, in an alkaline (pH9) to neutral (pH7) environment. But beware: Like their namesake, lobster mushrooms can develop an intense stink as they age.

Another famed dye mushroom is the Western jack-o'-lantern (*Omphalotus olivascens*), a species that glows in the dark. In a mildly acidic (pH5) environment, *O. olivascens* will create purple with the addition of alum mordant on wool fiber, or a deep green with iron mordant on wool and silk fibers. Even without a mordant, *O. olivascens* creates a lovely shade of purple that will shift to a lightfast reddish brown over time. *Gymnopilus ventricosus* is another species of white rot mushroom that thrives on living or dead wood and most likely contains hispidin, a source of yellow and orange pigment compounds (although it is unclear if hispidin is water soluble). For this species, an acidic (pH4) dye environment produces more brilliant yellows on wool fiber with the addition of tin mordant.

MUSHROOMS FEATURED IN THIS SECTION:

Cortinarius neosanguineus

Cortinarius ominosus

Cortinarius subcroceofolius

Cortinarius uliginosus

Gymnopilus ventricosus

Hypomyces lactifluorum

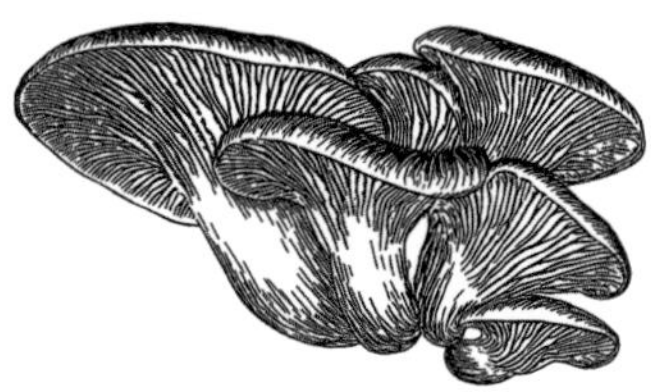

Omphalotus olivascens

Cortinarius neosanguineus

GENUS: *Cortinarius*

SPECIES: *neosanguineus*

COMMON NAME: Western Blood Red Cort

Cortinarius neosanguineus distinguishes itself from other *Cortinarius* in subgenus *Dermocybe* with its bright red color throughout its cap, gills, and stipe. It is fairly small, slender in stature, has a smooth surface, and when young a cobwebby veil is present. The pigment is concentrated in the cap but can also be found in the stipe and flesh. All parts can be used to obtain color.

HABITAT & ECOLOGY: Solitary or scattered on the ground, growing in moss or at the base of moss-covered logs and stumps. Found in conifer forests, specifically spruce and hemlock.

DISTRIBUTION: Northern North America

SPORE PRINT: Rusty red-brown

CAP: Bright red, turning more purple-red when wet. The rim around the edge appears lighter. Moist to dry, not sticky or slimy.

HYMENIUM: Gills broadly attached to the stipe. Bright red to purple-red, becoming rusty red-brown when spores mature.

STIPE: Long, thin, slender. Equal width throughout stipe but slightly larger at the base. Bright red to purple-red with fibrous flesh.

Dye Preparation

CONDITION	Dried
PART	Entire mushroom
PRE-SOAK	–
pH	10
RATIO	1:2
TEMP	110°F/43°C
TIME	1 hour

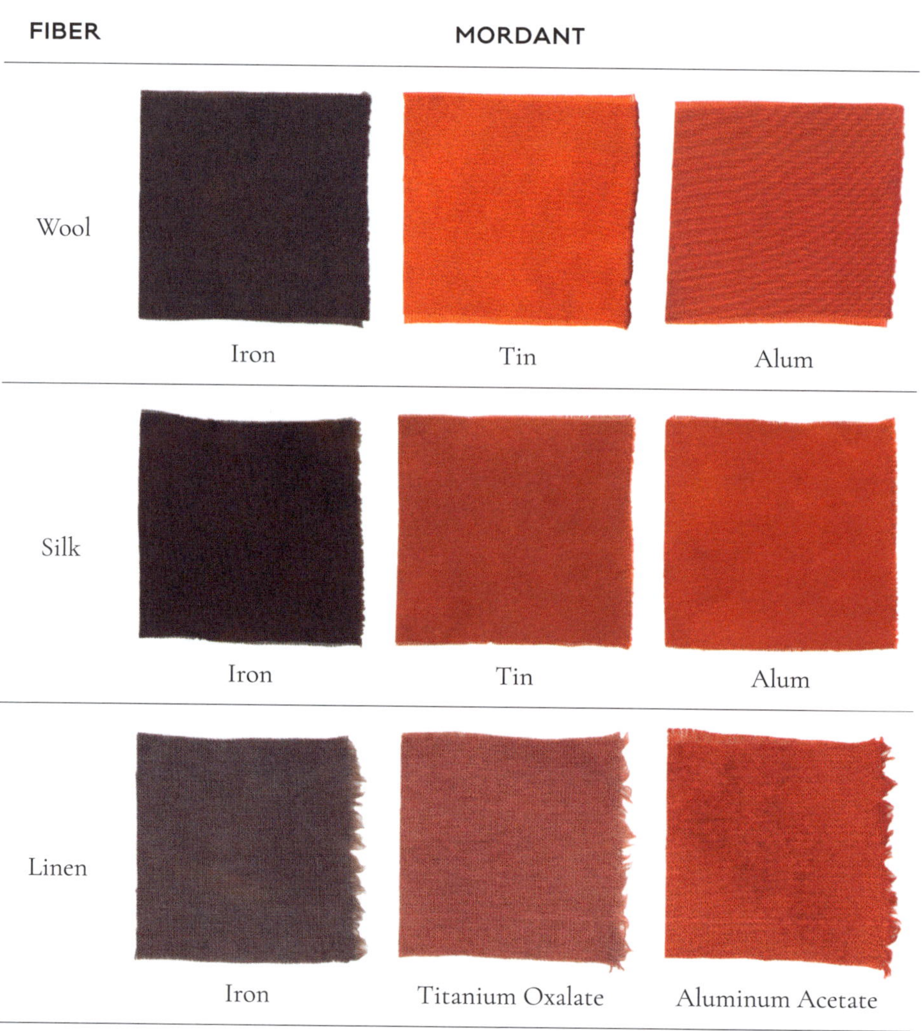

Pigment Preparation

CONDITION	Dried
PART	Entire mushroom
RINSE	Three times
BINDER	Gum Arabic
TEMP	85°F/29°C

PIGMENT

Dye Preparation

CONDITION	Dried
PART	Entire mushroom
PRE-SOAK	–
pH	7
RATIO	1:2
TEMP	110°F/43°C
TIME	1 hour

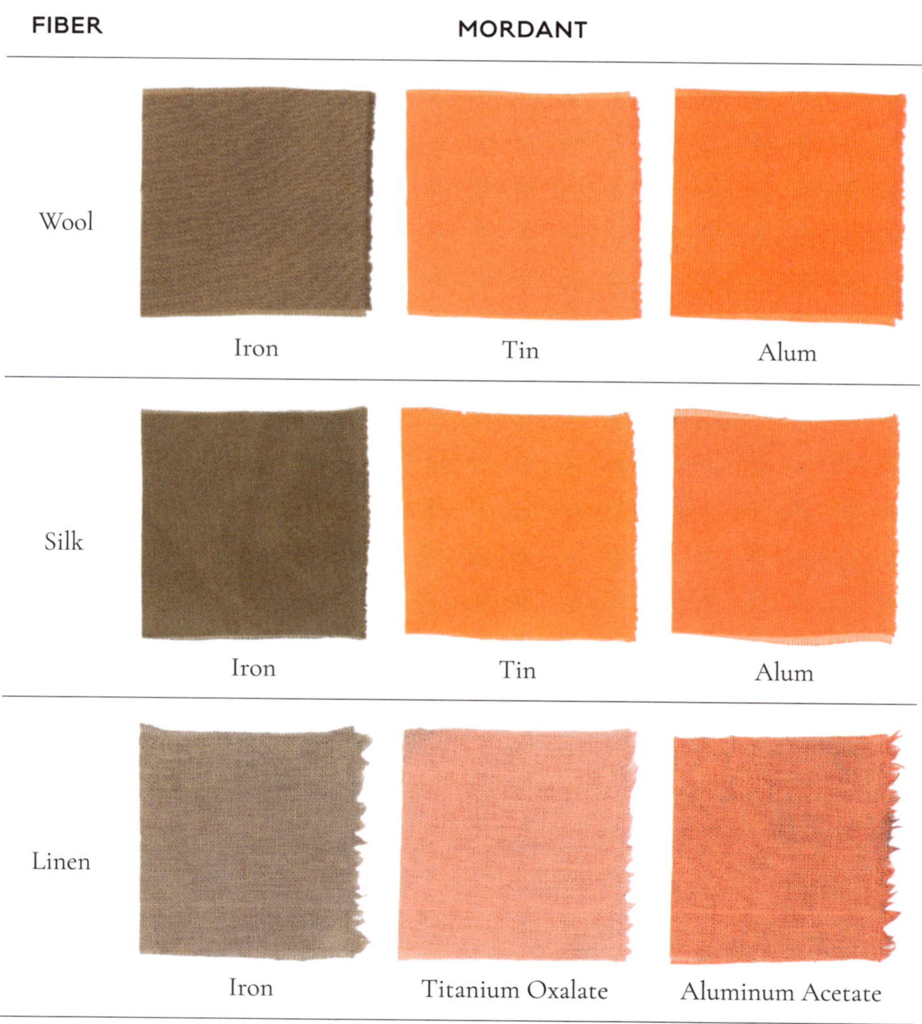

Pigment Preparation

CONDITION	Dried
PART	Entire mushroom
RINSE	Twice
BINDER	Gum Arabic
TEMP	85°F/29°C

PIGMENT

Base

Alum

Citric Acid

Iron

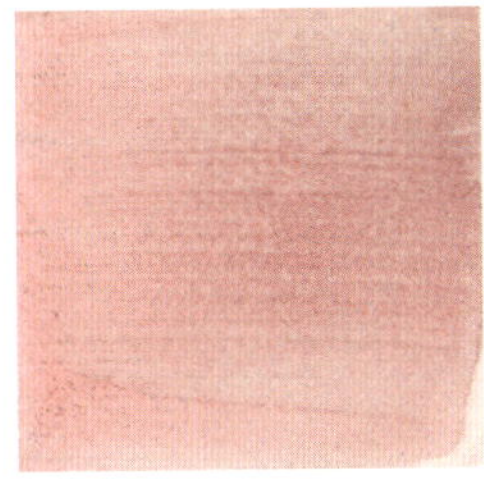

Copper Acetate

Soda Ash

Cortinarius ominosus

GENUS: *Cortinarius*

SPECIES: *ominosus*

COMMON NAME: Surprise Webcap

Cortinarius ominosus has a yellow to yellow-brown cap and stipe and red to dark red gills. It is fairly small, has a smooth surface, and when young a cobwebby veil is present. The red pigment is concentrated in the cap, with an orange pigment in its stipe. Caps and stipes are often separated before going into the dye pot to obtain two different colors. *Cortinarius semisanguineus* is a misapplied name.

HABITAT & ECOLOGY: Solitary or scattered on the ground, growing in moss or at the base of moss-covered logs and stumps. Primarily found in mixed conifer forests.

DISTRIBUTION: North America

SPORE PRINT: Rusty red-brown

CAP: Yellow to yellow-brown. The rim around the edge appears lighter. Moist to dry, not sticky or slimy.

HYMENIUM: Gills broadly attached to stipe. Red, becoming rusty red-brown when spores mature.

STIPE: Long and equal width throughout stipe but slightly larger at the base. Yellow to yellow-brown with fibrous flesh.

Dye Preparation

CONDITION	Fresh
PART	Caps
PRE-SOAK	–
pH	9
RATIO	1:1
TEMP	110°F/43°C
TIME	1 hour

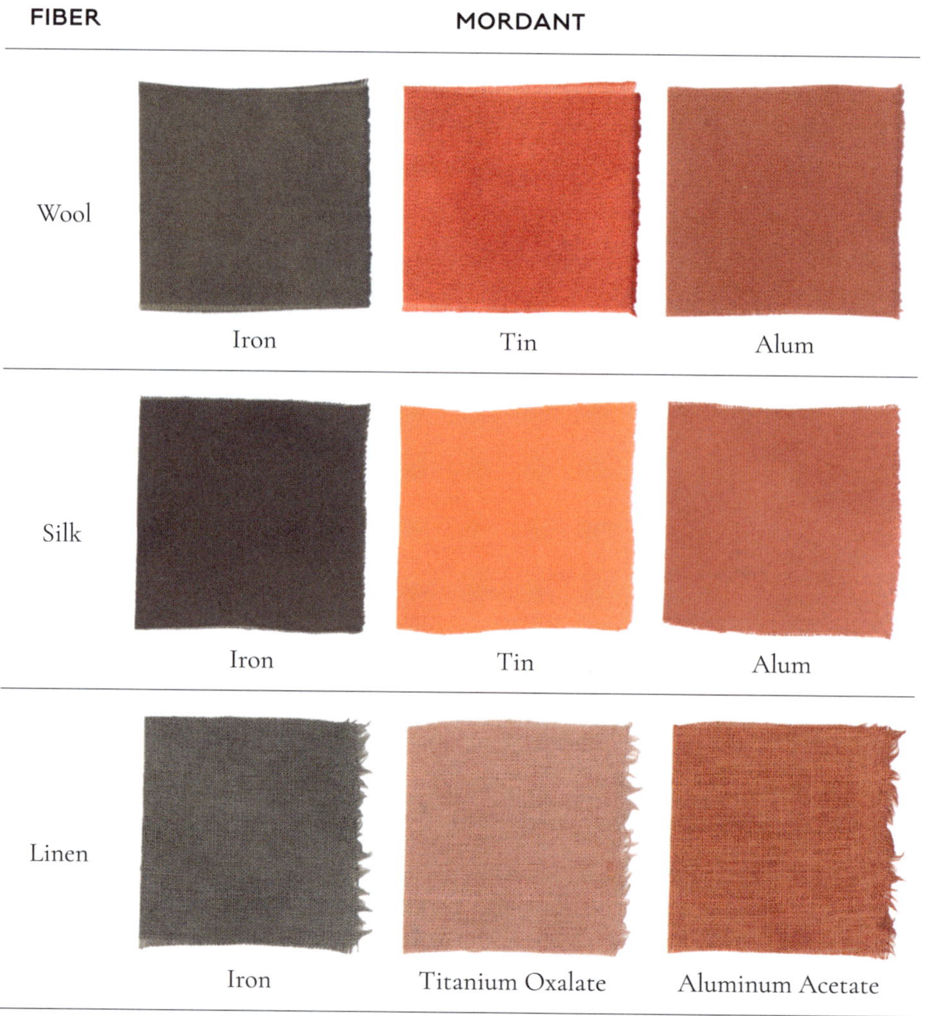

FIBER	MORDANT		
Wool	Iron	Tin	Alum
Silk	Iron	Tin	Alum
Linen	Iron	Titanium Oxalate	Aluminum Acetate

Pigment Preparation

CONDITION	Fresh
PART	Caps
RINSE	Three times
BINDER	Gum Arabic
TEMP	85°F/29°C

PIGMENT

Dye Preparation

CONDITION	Dried
PART	Caps
PRE-SOAK	–
pH	7
RATIO	1:1
TEMP	110°F/43°C
TIME	1 hour

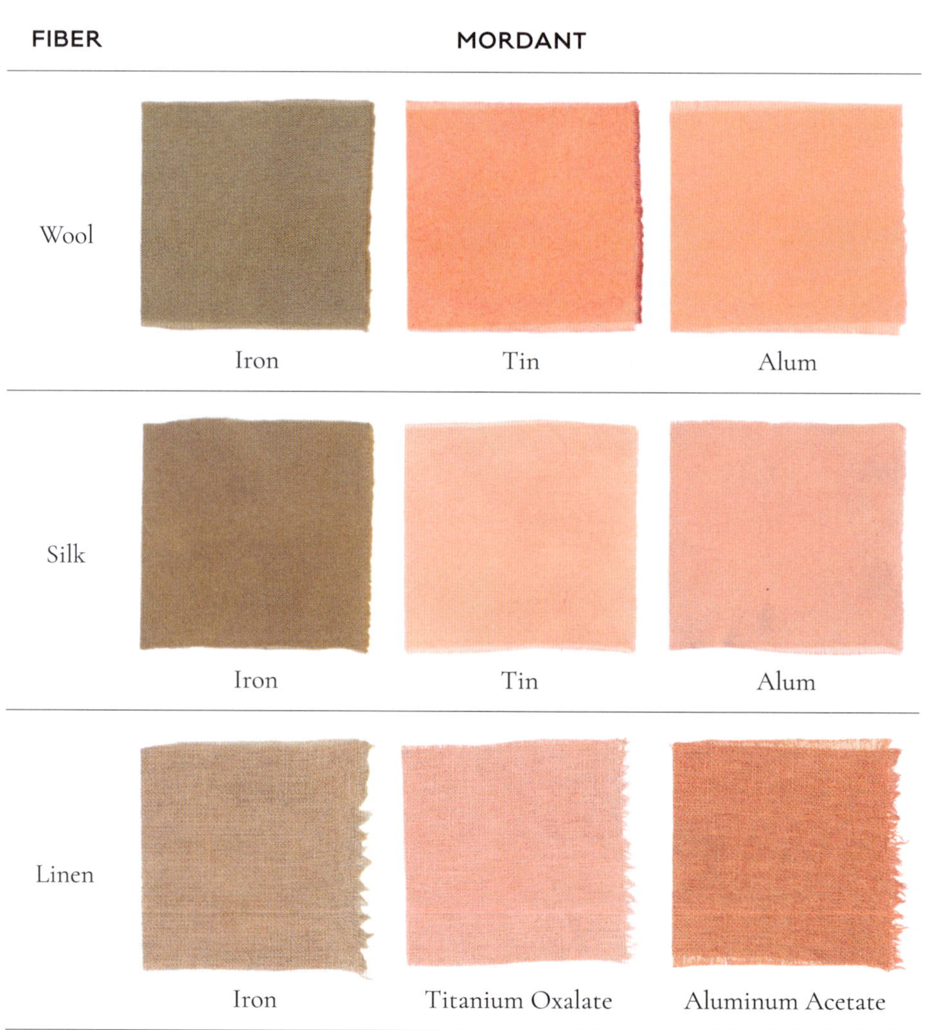

Pigment Preparation

CONDITION	Dried
PART	Caps
RINSE	Three times
BINDER	Gum Arabic
TEMP	85°F/29°C

PIGMENT

Base

Alum

Citric Acid

Iron

Copper Acetate

Soda Ash

Cortinarius subcroceofolius

GENUS: *Cortinarius*

SPECIES: *subcroceofolius*

COMMON NAME: Cinnamon Cort

Cortinarius subcroceofolius has a yellow-brown to orange-brown cap and stipe with bright yellow to orange gills. It is fairly small, has a smooth surface, and when young a cobwebby veil is present. The pigment is concentrated in the cap but can also be found in the stipe and flesh. All parts can be used to obtain color. It is also a synonym of *Cortinarius cinnamomeus*.

HABITAT & ECOLOGY: Solitary or scattered on the ground, growing in moss or at the base of moss-covered logs and stumps. Found in conifer forests, specifically pines.

DISTRIBUTION: Western North America

SPORE PRINT: Rusty red-brown

CAP: Light yellow-brown to dark orange-brown, turning more red-brown with age. The rim around the edge appears lighter. Moist to dry, not sticky or slimy.

HYMENIUM: Gills broadly attached to the stipe. Bright yellow when young, becoming more orange with age, and finally rusty orange-brown when spores mature.

STIPE: Long and equal width throughout stipe but slightly larger at the base. Light yellow-brown to brown-orange with fibrous flesh.

Dye Preparation

CONDITION	Dried
PART	Entire mushroom
PRE-SOAK	–
pH	5
RATIO	1:1
TEMP	110°F/43°C
TIME	1 hour

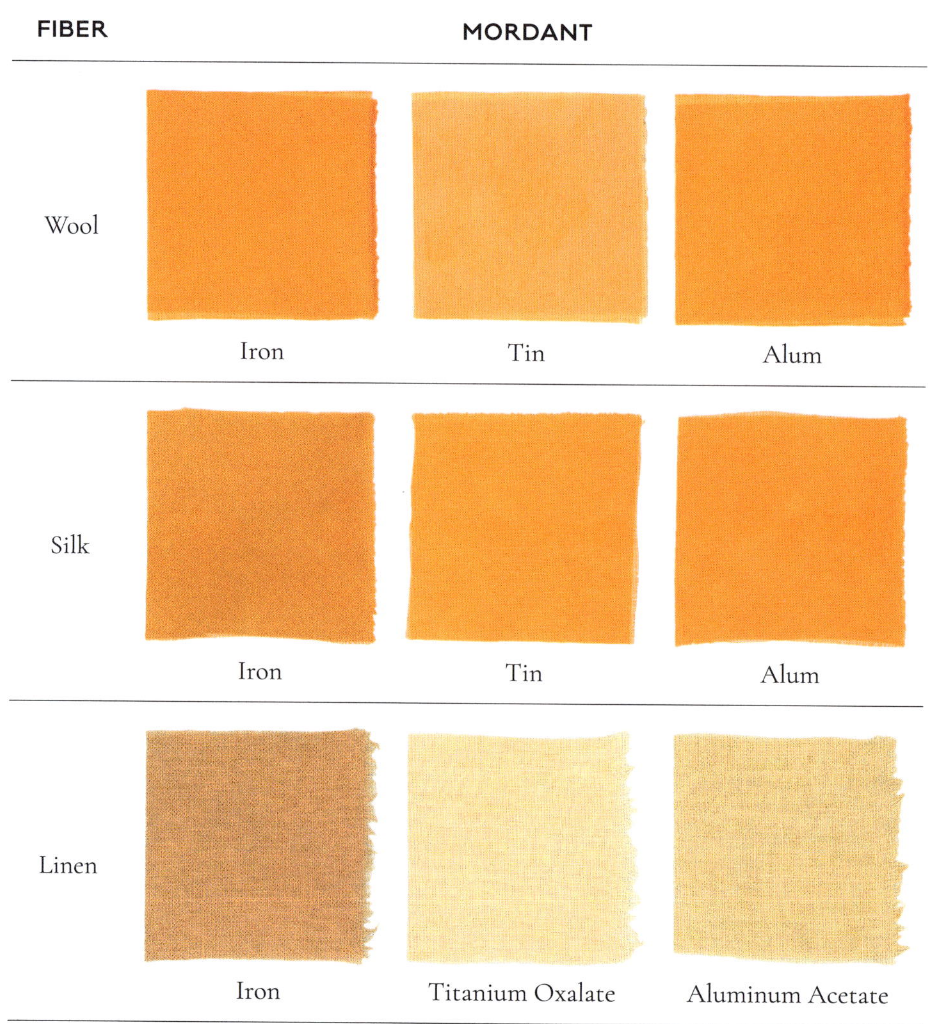

Pigment Preparation

CONDITION	Dried
PART	Entire mushroom
RINSE	Twice
BINDER	Gum Arabic
TEMP	85°F/29°C

PIGMENT

Base

Alum

Citric Acid

Iron

Copper Acetate

Soda Ash

Dye Preparation

CONDITION	Dried
PART	Entire mushroom
PRE-SOAK	–
pH	7
RATIO	1:1
TEMP	110°F/43°C
TIME	1 hour

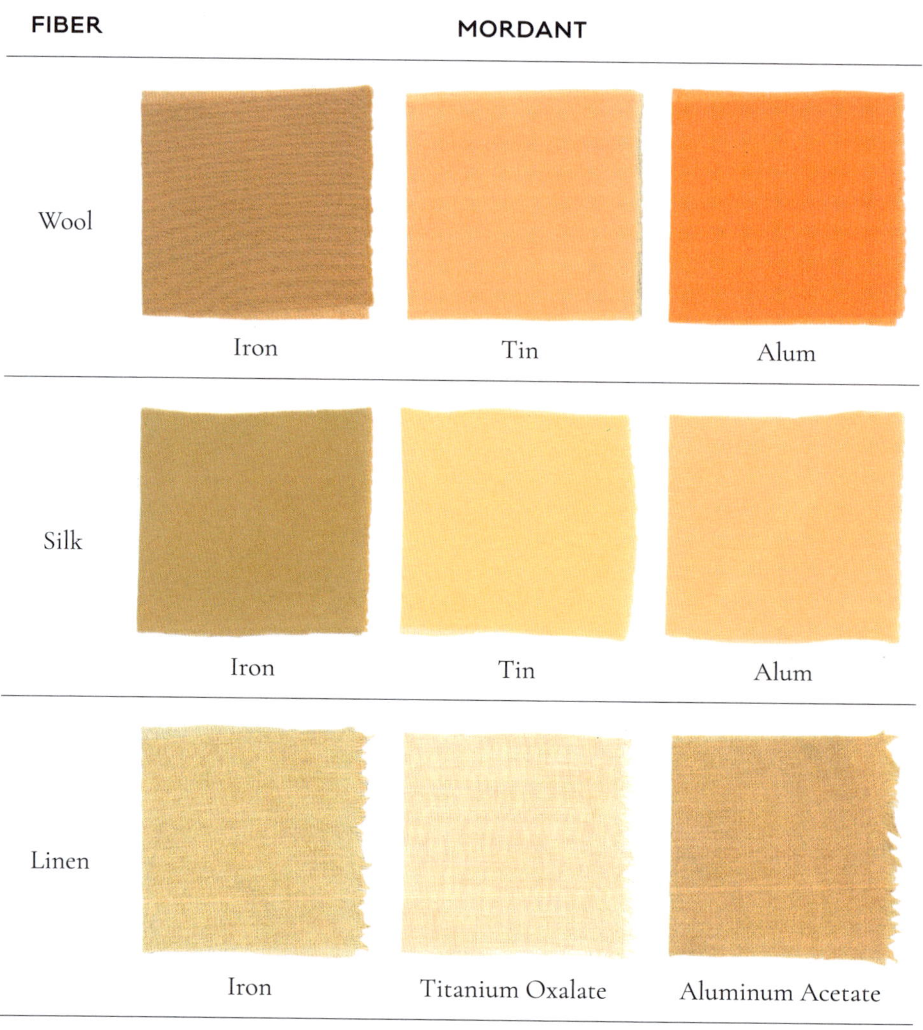

Pigment Preparation

CONDITION	Dried
PART	Entire mushroom
RINSE	Three times
BINDER	Gum Arabic
TEMP	85°F/29°C

PIGMENT

Base

Alum

Citric Acid

Iron

Copper Acetate

Soda Ash

Cortinarius uliginosus

GENUS: *Cortinarius*

SPECIES: *uliginosus*

COMMON NAME: Marsh Webcap

Cortinarius uliginosus has a brownish red-orange cap and gills and yellow-orange stipe. It is fairly small, has a smooth surface, and when young a cobwebby veil is present. The pink pigment is concentrated in the cap, with an orange pigment in its stipe. Caps and stipes are often separated before going into the dye pot to obtain two different colors.

HABITAT & ECOLOGY: Solitary or scattered, growing in damp soil in deciduous woods, often beside lakes, rivers, bogs, and marshes. Prefers willow and alder trees.

DISTRIBUTION: North America, United Kingdom

SPORE PRINT: Rusty red-brown

CAP: Brownish red-orange. The rim around the edge appears lighter, becoming upturned and wavy with age.

HYMENIUM: Gills broadly attached to the stipe. Bright yellow when young, turning yellow-orange in color with age, and finally rusty red-brown when spores mature.

STIPE: Long and equal width throughout the stipe with a slightly swollen base. Yellow with fibrous flesh.

Dye Preparation

CONDITION	Dried
PART	Caps
PRE-SOAK	–
pH	7
RATIO	1:1
TEMP	110°F/43°C
TIME	1 hour

Pigment Preparation

CONDITION	Dried
PART	Caps
RINSE	Three times
BINDER	Gum Arabic
TEMP	85°F/29°C

PIGMENT

Gymnopilus ventricosus

GENUS: *Gymnopilus*

SPECIES: *ventricosus*

COMMON NAME:
Western Jumbo Gym

Gymnopilus ventricosus is a large orange-yellow mushroom that fruits in clusters on or near decaying wood. It is virtually indistinguishable from *Gymnopilus voitkii*, which has larger spores. Bright yellow-orange gills under the cap and a light yellow stipe, often displaying a skirt (remnants of the partial veil), make this pair of look-alikes easy to spot in the forest. The pigment is contained throughout the entire mushroom, with higher concentrations in the cap.

HABITAT & ECOLOGY: Almost always in large clusters, growing on or near the base of conifer or deciduous decaying trees or stumps, seldom high up on the trunk.

DISTRIBUTION: North America

SPORE PRINT: Rusty orange-brown

CAP: Yellow to orange with streaks of darker red-brown with age. Surface can have remnants of partial veil but is often smooth, dry to moist.

HYMENIUM: Notched gills attached to the stipe. Light yellow becomes bright red-brown and dark brown with age. Distinct edges, close and irregular shaped gills.

STIPE: Long, thick, curved, and swollen in the lower section, tapering at the base. Light yellow with red-brown vertical streaks.

Dye Preparation

CONDITION	Dried
PART	Entire mushroom
PRE-SOAK	–
pH	4
RATIO	1:1
TEMP	160°F/71°C
TIME	1 hour

Pigment Preparation

CONDITION	Dried
PART	Entire mushroom
RINSE	Three times
BINDER	Gum Arabic
TEMP	85°F/29°C

PIGMENT

Base

Alum

Citric Acid

Iron

Copper Acetate

Soda Ash

Hypomyces lactifluorum

GENUS: *Hypomyces*

SPECIES: *lactifluorum*

COMMON NAME: Lobster Mushroom

Hypomyces lactifluorum is a parasitic fungus that most often attacks and grows on the host *Russula brevipes* mushroom (and occasionally other species), creating a bright red-orange crust. Pigment can be found in the outer layer and peeled off (discarding the white flesh) for use in the dye pot. It maintains the shape of the *R. brevipes* but with contorting of the gills and an orange to red grainy texture, making it a uniquely distinctive mushroom. Collecting the oldest, most mature red-purple specimens yields the most intense pigment, but beware: They are smelly!

HABITAT & ECOLOGY: Parasitic on *R. brevipes*, found on the ground, buried in duff before maturity. Solitary, scattered, or clustered.

DISTRIBUTION: Asia, Australia, Europe, North America

SPORE PRINT: White

CAP: Light orange-red to bright orange-red. Red and purple stains with age. Surface is dry to moist, often pitted or cracked.

HYMENIUM: Gills are covered with parasitic fungus, turning them into ridges, often vertical and upright.

STIPE: Often distorted, short, and thick, but sometimes more fully formed.

Dye Preparation

CONDITION	Fresh
PART	Entire mushroom
PRE-SOAK	–
pH	9
RATIO	2:1
TEMP	145°F/63°C
TIME	1 hour

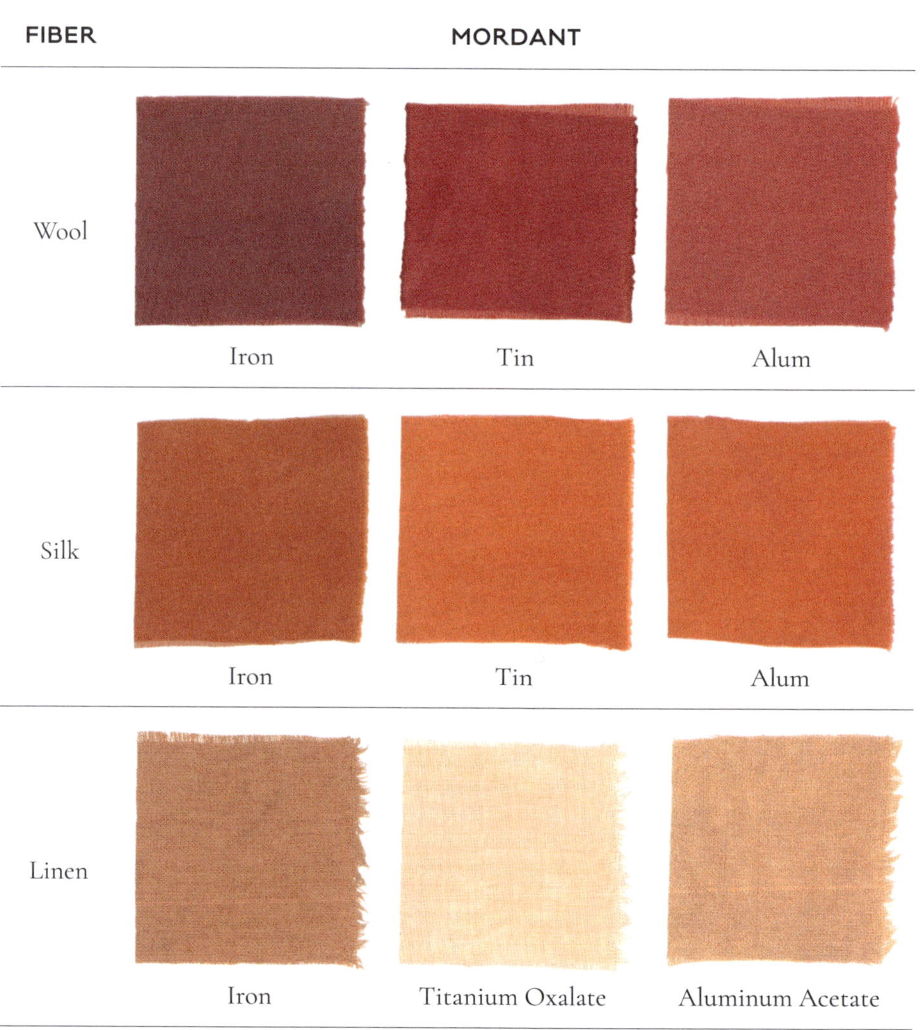

Pigment Preparation

CONDITION Fresh
PART Entire mushroom
RINSE Three times
BINDER Gum Arabic
TEMP 85°F/29°C

PIGMENT

Dye Preparation

CONDITION Dried
PART Entire mushroom
PRE-SOAK –
pH 10
RATIO 2:1
TEMP 145°F/63°C
TIME 1 hour

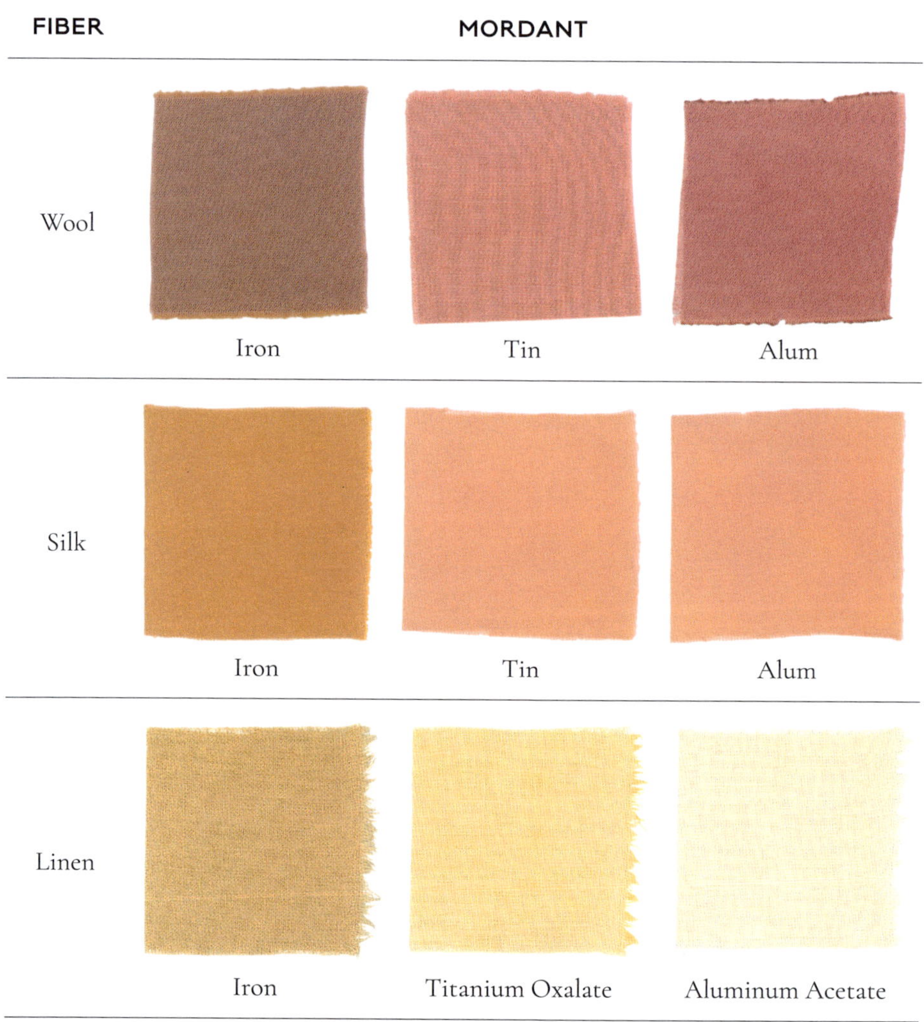

Pigment Preparation

CONDITION	Dried
PART	Entire mushroom
RINSE	Twice
BINDER	Gum Arabic
TEMP	85°F/29°C

PIGMENT

Base

Alum

Citric Acid

Iron

Copper Acetate

Soda Ash

Omphalotus olivascens

GENUS: *Omphalotus*

SPECIES: *olivascens*

COMMON NAME: Western Jack-o'-Lantern

Omphalotus olivascens is a bright yellow-orange to olive-brown mushroom. Most often growing in overlapping (sometimes huge!) clusters on stumps, roots, and logs. This unique, bioluminescent mushroom is only found in California and a small section of Northern Mexico. It can yield purple and green hues.

HABITAT & ECOLOGY: Large, overlapping clusters, sometimes solitary or in small clusters, growing on hardwood stumps, roots, and logs, specifically oak.

DISTRIBUTION: California, Northern Mexico

SPORE PRINT: White to pale yellow

CAP: Yellow-orange to olive-brown, uplifting and wavy with age. Easily watermarked with olive-green stains, smooth surface, becoming cracked when dried out.

HYMENIUM: Bright yellow-orange to dingy olive-yellow, paler when covered in white spores. Decurrent gills running down stipe.

STIPE: Center to off-center, thick, curved base penetrating deep into wood substrate. Smooth to rough surface.

Dye Preparation

CONDITION	Dried
PART	Entire mushroom
PRE-SOAK	–
pH	5
RATIO	1:1
TEMP	140°F/60°C
TIME	1 hour

Pigment Preparation

CONDITION	Dried
PART	Entire mushroom
RINSE	Twice
BINDER	Gum Arabic
TEMP	85°F/29°C

PIGMENT

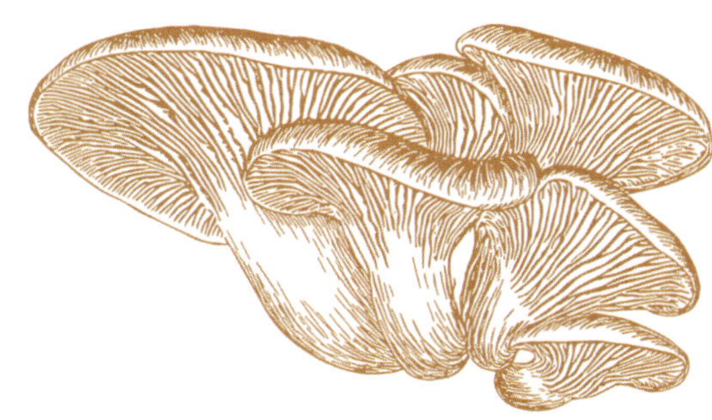

CHAPTER THREE

Polypores

More than a decade ago, when I first searched online to learn if mushrooms could be used for color, I immediately came across pictures of the famed dyer's polypore (*Phaeolus schweinitzii*). I pored over the images of creamy white-rimmed glowing orange-brown rosettes. Without the true stipe or cap of a traditional-looking mushroom, this species defied my understanding of what a mushroom should look like. Its shape was flowerlike, it looked fibrous and thick, and I was surprised to see it often had plants growing out of it. I would come to learn this type of growth is considered "indeterminate"—the mushroom just grows up and around anything in its way. I made this the first dye mushroom on my must-find list and hit the trails. It took me almost a full season, but when I spotted *P. schweinitzii* in the forest, I was struck by the shimmering yellow, orange, and cream rings around its edges. Turning it over, I could see the same colors on its spore-bearing surface as well as a deep green color. I did a happy myco dance and immediately brought it back to my studio to check its identification. I began to draw it in my sketchbook, abstracting the shapes of the rosettes, removing debris, examining the plants growing through it, and simply admiring my first dyer's polypore. When I dared to cook it up and extract its colors, the magical results did not disappoint.

This large group of fungi, the polypores, includes a wide array of species that all have a layer of thousands of tiny tubes on the underside of the cap. These small cylindrical tubes release spores out of the open ends, called pores. In some cases, these surfaces are so smooth, irregular, or wrinkled that it's hard to see the pored layer, but if you look closely, especially with a field lens or microscope, the subtle spongelike pattern of pores will be revealed.

These tough, bracket-like fungi are major decayers of wood. Most polypores are saprobes (living on decaying organisms), although some are ectomycorrhizal and live in cooperation with the roots of living trees. The shapes, sizes, textures, and growth habits of these mushrooms are all over the place! Some are very tough, woody, and leathery, but others are watery and soft. Some look like a typical mushroom with a stipe and cap—these are known as stalked polypores, such as *Boletopsis grisea*. Polypores can also form shelves, or conks, growing off the sides of decaying logs, stumps, or trees, and can range from hoof-shaped to fan-shaped. They may also fruit in large clusters with overlapping tiers like the porous, spongy *Pycnoporellus alboluteus*. Not many polypores are edible, but some possess great medicinal potential.

The range of hues produced by polypore mushrooms is extreme, from Easter egg purple to golden yellow, to an intense vibrant orange, a deep forest green, and everything in between. These mushrooms are often able to break down lignin, the resinous substance in wood, producing white rot. This process yields cinnamic acid as well as other substances such as hispidin and hypholomin, each a source of yellow and orange pigment compounds.[1] Polypores such as *Hapalopilus nidulans* contain polyporic acid, which produces a heady purple in combination with wool fibers (though, curiously, it doesn't perform well on silk fibers). Often, to get the richest color from them, each species of polypore needs to be treated in a particular way. *H. nidulans* performs well in a neutral (pH7) dye bath and

---

1. Miriam C. Rice, illustrated by Dorothy Beebee, *Mushrooms for Color* (Eureka: Mad River Press, 1980), 147.

achieves darker colors in an alkaline (pH9) solution, while *P. schweinitzii* will produce a range of colors in both an acidic (pH5) as well as an alkaline (pH9) environment, resulting in golden yellows, vibrant oranges, forest greens, and rust browns. It will produce a lovely yellow on unmordanted fiber, but adding mordants will result in a richer, deeper golden yellow. *B. grisea* performs best in an alkaline (pH9) dye bath and yields a range of shades of green. *P. alboluteus* makes a variety of golden hues in an acidic (pH4) environment, and the pigment will transform into a pink-red with an alkaline (pH9) soda ash modifier.

MUSHROOMS FEATURED IN THIS SECTION:

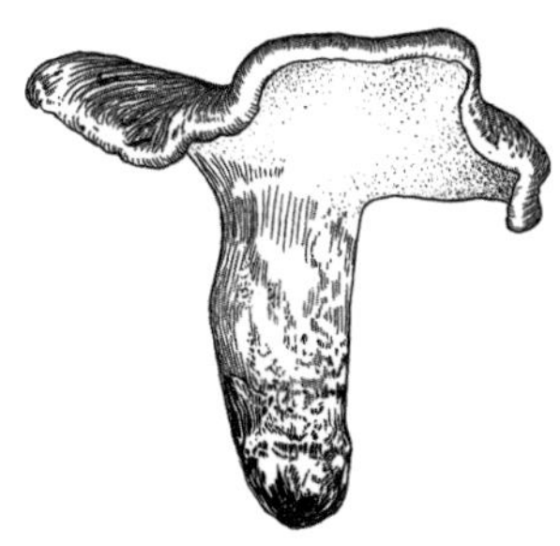

Boletopsis grisea

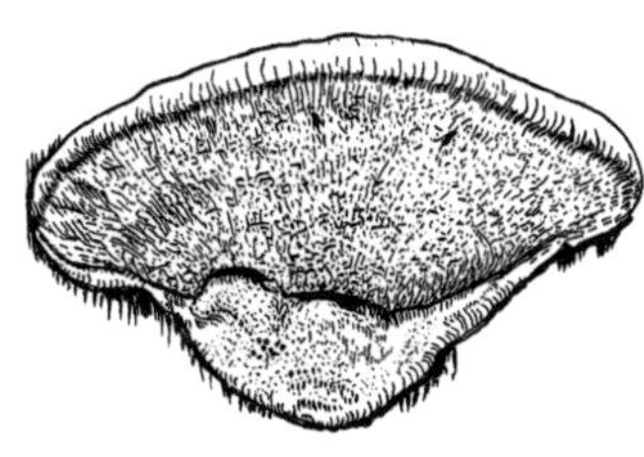

Hapalopilus nidulans

Phaeolus schweinitzii

Pycnoporellus alboluteus

Boletopsis grisea

GENUS: *Boletopsis*

SPECIES: *grisea*

COMMON NAME:
Gray False Bolete

Boletopsis grisea is categorized as a stalked polypore. It is light gray, developing dark gray streaks with age. It is smooth with firm flesh, becoming brittle in time. Often hidden in the duff of pine needles (disguised as rocks), these mushrooms stain green to gray-green with KOH.

HABITAT & ECOLOGY: Solitary or scattered, growing on the ground with pines, often hidden under duff.

DISTRIBUTION: Europe, North America

SPORE PRINT: Light brown

CAP: Off-white to light gray when young, developing dark gray streaks and spots at maturity. Smooth, dry surface forming cracks when dried out.

HYMENIUM: Smooth, shallow pores that easily rub off, white to light gray.

STIPE: Thick, equal width throughout with a rounded base. Light gray developing dark gray streaks with age. Dry, smooth surface.

Dye Preparation

CONDITION	Dried
PART	Entire mushroom
PRE-SOAK	–
pH	9
RATIO	1:2
TEMP	165°F/74°C
TIME	1 hour

FIBER	MORDANT		
Wool	Iron	Tin	Alum
Silk	Iron	Tin	Alum
Linen	Iron	Titanium Oxalate	Aluminum Acetate

Pigment Preparation

CONDITION	Dried
PART	Entire mushroom
RINSE	Twice
BINDER	Gum Arabic
TEMP	85°F/29°C

PIGMENT

Base

Alum

Citric Acid

Iron

Copper Acetate

Soda Ash

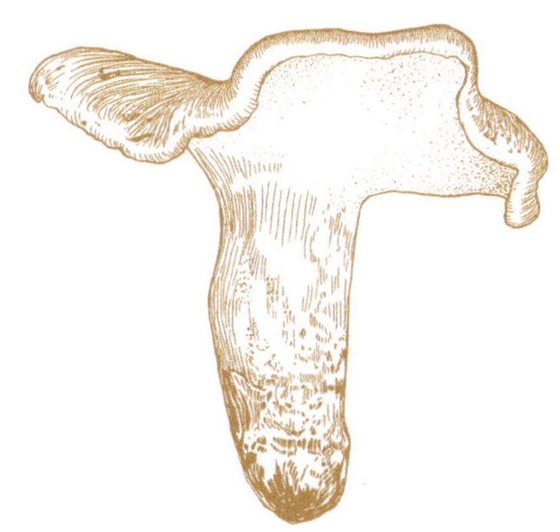

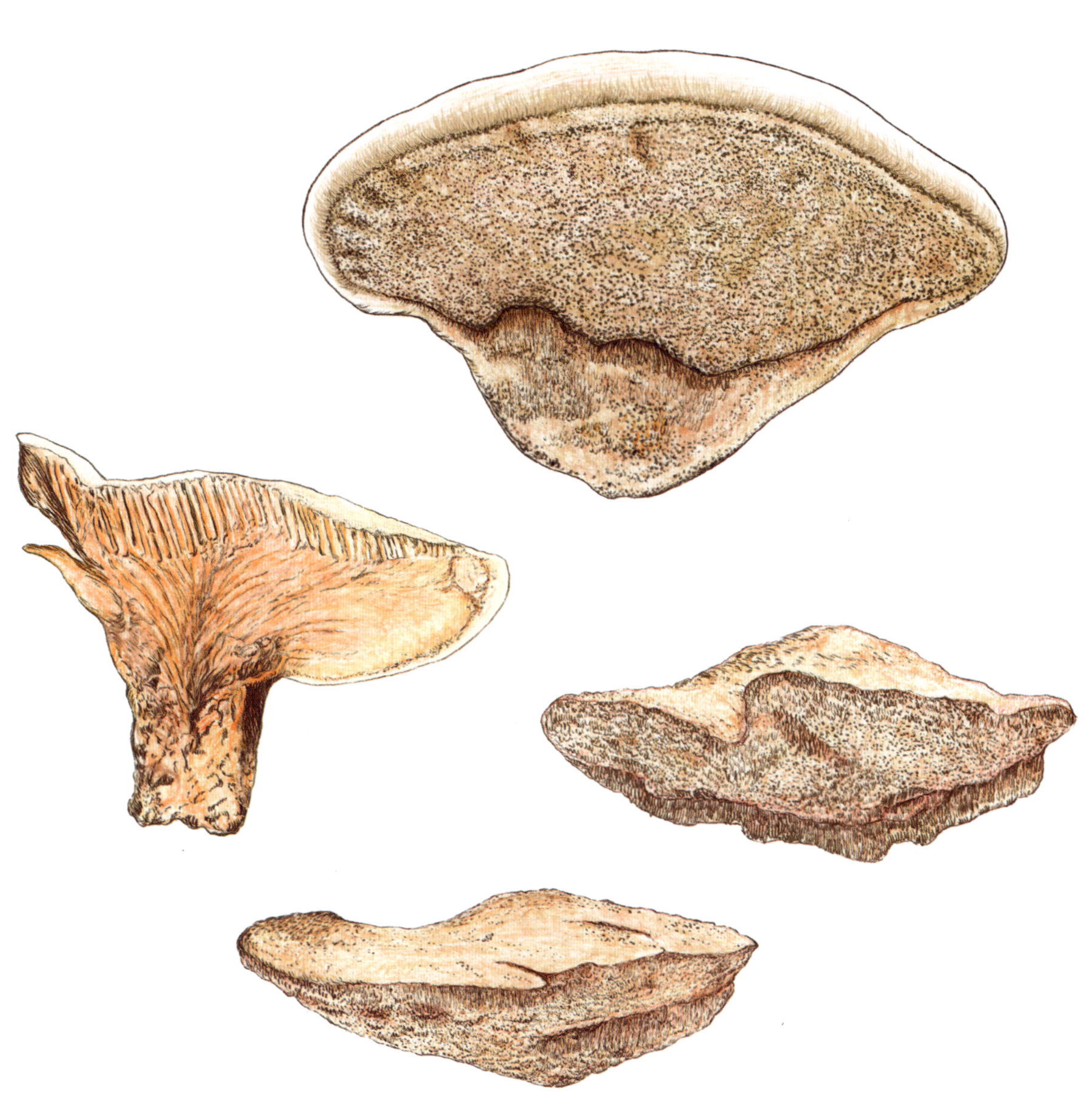

Hapalopilus nidulans

GENUS: *Hapalopilus*

SPECIES: *nidulans*

COMMON NAME: Cinnamon Bracket

Hapalopilus nidulans is a kidney-shaped, fleshy beige-orange-brown polypore growing on decaying logs or branches. This highly prized mushroom stains bright purple with KOH. The entire mushroom can be used to obtain a lilac-purple color.

HABITAT & ECOLOGY: Solitary or scattered, overlapping clusters on deciduous decaying logs or branches, such as oak, beech, and birch.

DISTRIBUTION: Asia, Australia, Eastern North America, Europe, North Africa

SPORE PRINT: White

CAP: Fleshy and soft, beige orange-brown, fan shaped to semicircular. Covered with matted hairs. Becoming brown-orange, brittle, and hard with age.

HYMENIUM: Yellow-brown with tiny angular pores.

STIPE: Lacks a stipe in the true sense; instead attaches to the side of decaying logs and branches.

Dye Preparation

CONDITION Dried
PART Entire mushroom
PRE-SOAK Yes
pH 7
RATIO 1:3
TEMP 145°F/63°C
TIME 1 hour

FIBER	MORDANT		
Wool	Iron	Tin	Alum
Silk	Iron	Tin	Alum
Linen	Iron	Titanium Oxalate	Aluminum Acetate

Pigment Preparation

CONDITION	Dried
PART	Entire mushroom
RINSE	Once
BINDER	Gum Arabic
TEMP	85°F/29°C

PIGMENT

Base

Alum

Citric Acid

Iron

Copper Acetate

Soda Ash

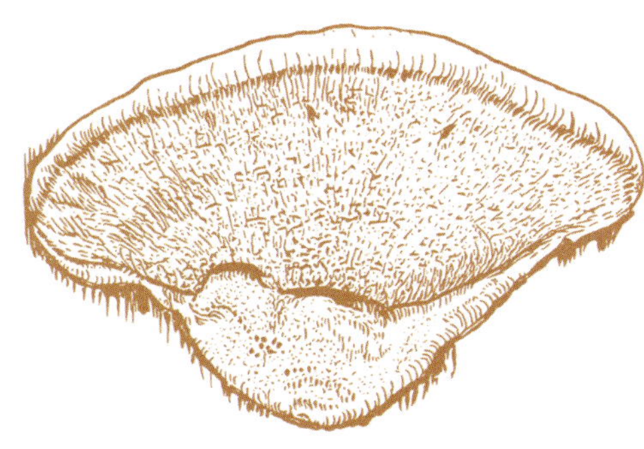

Phaeolus schweinitzii

GENUS: *Phaeolus*

SPECIES: *schweinitzii*

COMMON NAME: Dyer's Polypore

Phaeolus schweinitzii forms an irregularly shaped blob, eventually expanding into a circular shape and ultimately into a rosette with multiple shelves. It has a dark center with a bright yellow-green rim around its edge when mature and becomes entirely dark red-brown with age. The underside is spongy to touch, bright yellow to olive green when mature. Pigment is contained in mature specimens and not present in old, dark brown specimens from previous seasons.

HABITAT & ECOLOGY: Solitary or scattered, growing on the ground at the base of dead and dying conifers. Occasionally shelflike on stumps.

DISTRIBUTION: Africa, Asia, Australia, Europe, New Zealand, North America

SPORE PRINT: White

CAP: A dry, velvety surface becoming bumpy with age. Irregular in shape, becoming a circular rosette with age. The color ranges from bright yellow with a pinkish-cream rim to a dark center with a bright yellow-green rim when mature, to a dark red-brown with dark yellow-brown rim when older.

HYMENIUM: A shallow sponge layer with a pore surface that is yellow-green when mature and rusty red-brown when old.

STIPE: Short, stout, irregular stipe forming from a narrow base.

Dye Preparation

CONDITION	Dried
PART	Entire mushroom
PRE-SOAK	–
pH	5
RATIO	1:2
TEMP	160°F/71°C
TIME	1 hour

FIBER	MORDANT		
Wool	Iron	Tin	Alum
Silk	Iron	Tin	Alum
Linen	Iron	Titanium Oxalate	Aluminum Acetate

Pigment Preparation

CONDITION	Dried
PART	Entire mushroom
RINSE	Twice
BINDER	Gum Arabic
TEMP	85°F/29°C

PIGMENT

Base

Alum

Citric Acid

Iron

Copper Acetate

Soda Ash

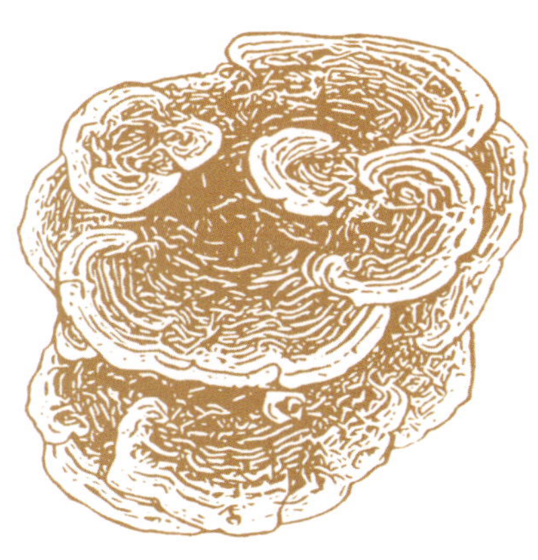

Dye Preparation

CONDITION	Dried
PART	Entire mushroom
PRE-SOAK	–
pH	9
RATIO	1:2
TEMP	160°F/71°C
TIME	1 hour

Pigment Preparation

CONDITION	Dried
PART	Entire mushroom
RINSE	Twice
BINDER	Gum Arabic
TEMP	85°F/29°C

PIGMENT

Base

Alum

Citric Acid

Iron

Copper Acetate

Soda Ash

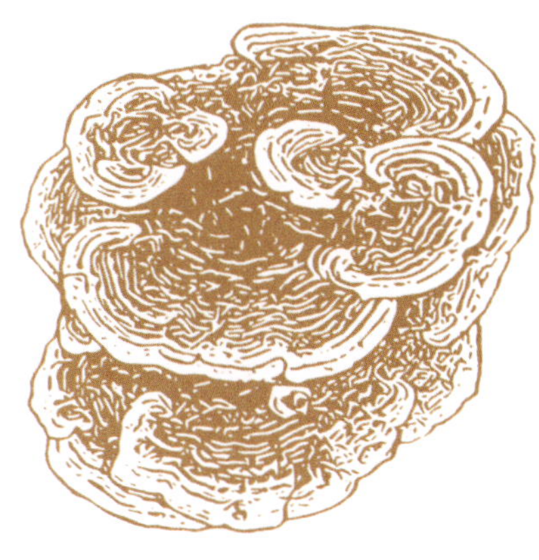

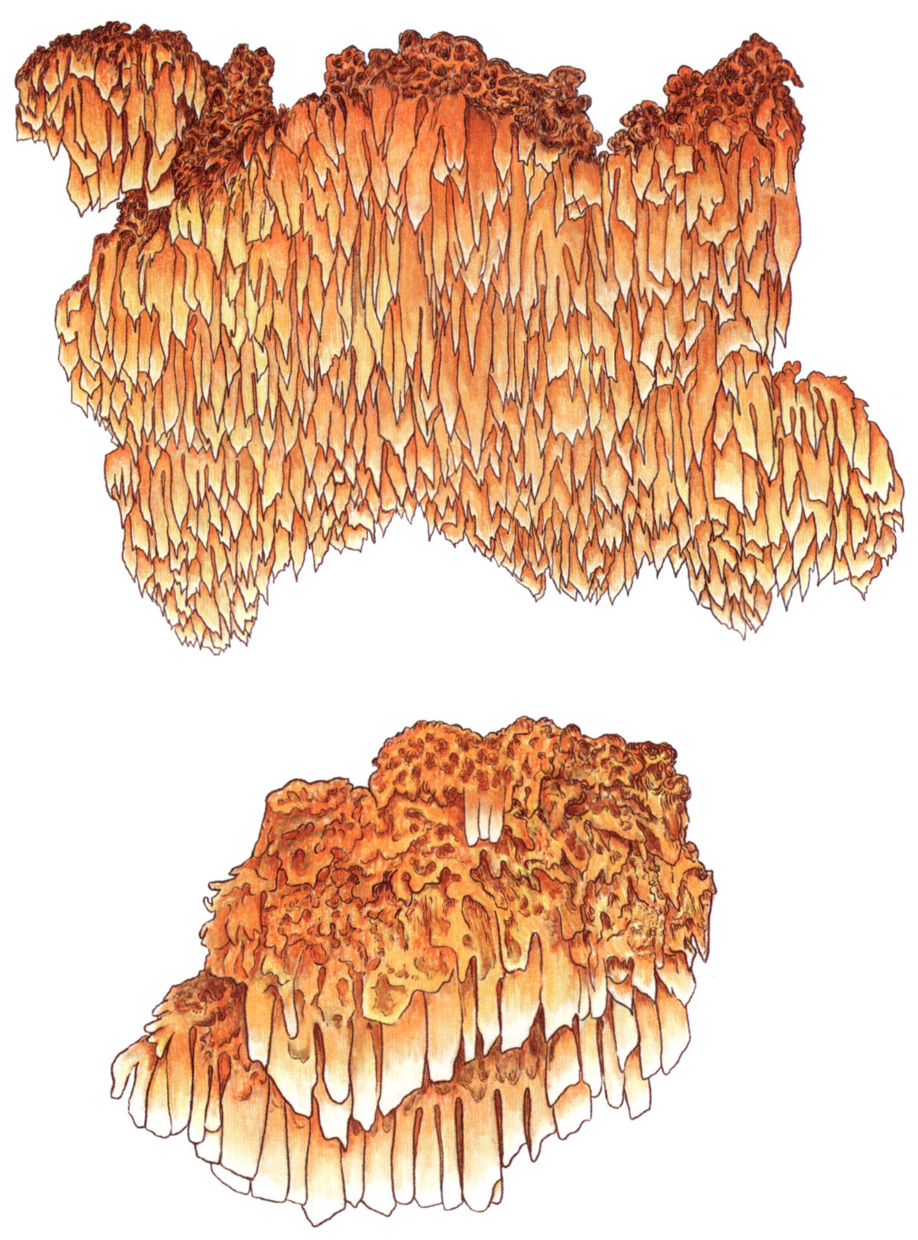

Pycnoporellus alboluteus

GENUS: *Pycnoporellus*

SPECIES: *alboluteus*

COMMON NAME:
Orange Sponge Polypore

Pycnoporellus alboluteus is a bright red-orange to light orange to white spongy spreading mushroom. It is irregular in shape, and its large pores can erode to look more like teeth. It grows on the lower side of decaying logs and stains bright red-orange with KOH.

HABITAT & ECOLOGY: A high elevation mountain fungi spreading up to 3 feet (approximately 1 meter), growing on the lower side of decaying logs under snowbanks, specifically spruce, firs, hemlocks, and occasionally aspen.

DISTRIBUTION: North America

SPORE PRINT: White

CAP: Soft, spongy, easy to peel off from the surface of wood.

HYMENIUM: Large, angular pores can erode to look more like teeth. Flesh is thin and bright orange to light orange.

STIPE: Lacks a stipe in a true sense; instead attaches to the side of a decaying log.

Dye Preparation

CONDITION	Dried
PART	Entire mushroom
PRE-SOAK	–
pH	4
RATIO	2:1
TEMP	145°F/63°C
TIME	1 hour

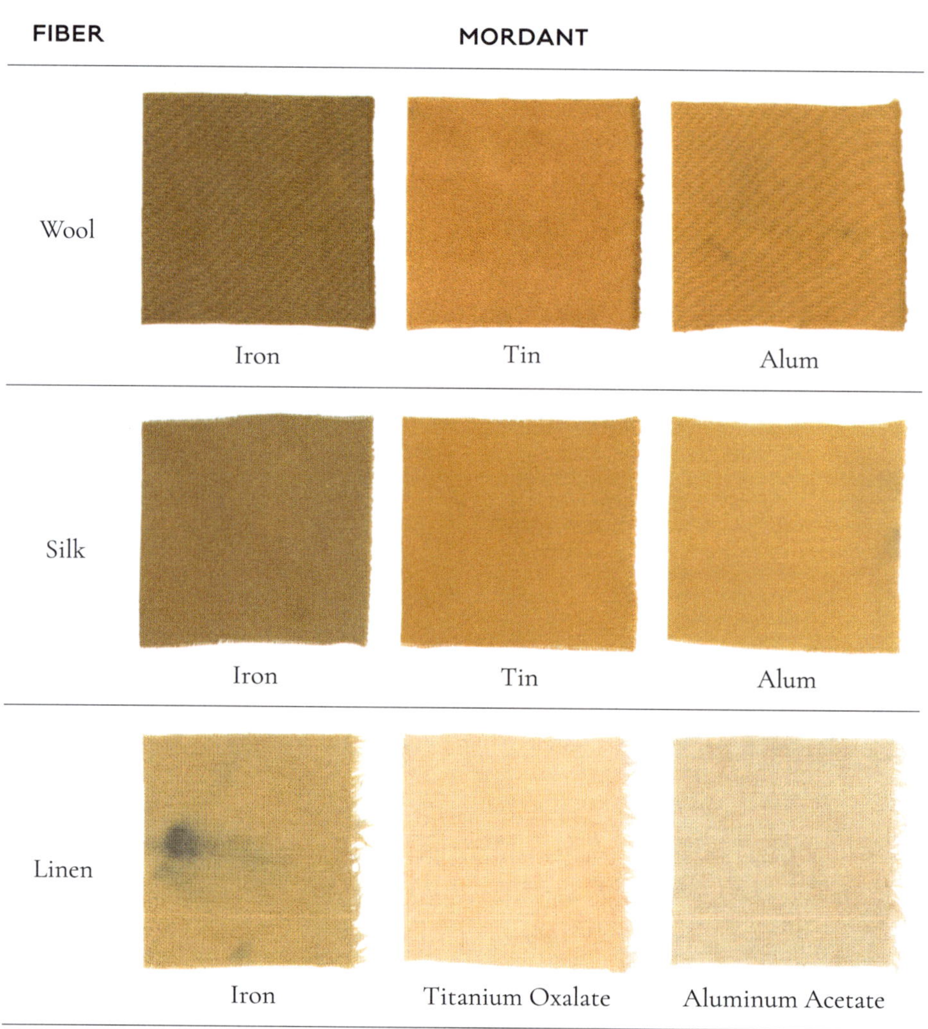

Pigment Preparation

CONDITION	Dried
PART	Entire mushroom
RINSE	Three times
BINDER	Gum Arabic
TEMP	85°F/29°C

PIGMENT

Base

Alum

Citric Acid

Iron

Copper Acetate

Soda Ash

CHAPTER FOUR

Tooth

Tooth fungi are a fascinating group; they produce spores on icicle-, stalactite-, and toothlike spines. Barring a few exceptions, like *Echinodontium tinctorium*, most tooth fungi have a stipe and cap. When I first learned about "tooth fungi," my imagination ran wild, and I quickly added the bleeding tooth mushroom (*Hydnellum peckii*) to the top of my must-find list. I'd studied this species in detail and heard encouraging stories of people finding it close to where I live. That autumn, I came across what I believed was a bleeding tooth mushroom literally gushing on the forest floor. It had a rosette shape, it was blueish, grayish, and whitish in color, and it had the classic red guttation (mushroom sweat!) oozing out of it. Confident I had found a real bleeding tooth mushroom, I took it home to confirm my identification. I sliced open the tough woody mushroom and something didn't check out; it was dark blue and orange on the inside and none of my field guides noted this. I had to do a deep dive into the ID characteristics of this mushroom to discover that what I had was *Hydnellum caeruleum*! While studying the *Hydnellum* genus, I learned these are coveted dye mushrooms that create a calming blue-green color. I was happy about this serendipitous find and even happier when, a short time later, I did find the real deal, *H. peckii*, near where I live.

The *Hydnellum*, *Phellodon*, and *Sarcodon* genera of tooth fungi contain unique pigmented chemical compounds found nowhere in nature except in the fungi kingdom. These compounds, known as terphenylquinones, are capable of producing elusive blue hues and occur in tooth fungi, polypores, and boletes. But achieving convincing blue and green colors from mushrooms is not as simple as achieving most other colors.

With the species *H. caeruleum*, *H. fuscoindicum*, *H. regium*, *S. squamosus*, *H. suaveolens*, and *P. niger*, I use a couple of different approaches to extract blue colors. One method is to modify the dye bath to an alkaline (pH9) solution within the first five to ten minutes of cooking the mushrooms and maintain an alklaline (pH9) dye bath the entire time they are cooking. The use of mordants such as alum, titanium oxalate, aluminum acetate with tannins in combination with cellulose fibers, and iron improves the overall color and lightfastness. The temperature of the dye bath is critical: The blue pigment will start to break down above 170°F (77°C) but needs to be at least 150°F (66°C) in order for the mushroom to release the pigment. An alternative method that yields good results is to pre-soak the mushrooms for days, weeks, or longer in an alkaline (pH9) solution before cooking them. These mushrooms also perform beautifully on cellulose fibers since cellulose fibers love an alkaline environment like the plant dye indigo. The sky blue of *H. caeruleum* on linen fiber is reminiscent of the soft blues obtained from indigo.

One critical determining factor is the quantity of mushrooms needed to achieve a deep, dark blue. The percentage of blue pigment in some mushrooms is quite low, often comprising less than 5% of the total dried weight.[1] Increasing the ratio of mushrooms to fiber, sometimes as much as ten times, will yield deeper shades of blue. This is especially true when working with species such as *H. caeruleum* and *H. suaveolens*. However, working with *H. regium* and *P. niger* yields substantial blues with only the same amount of mushroom to fiber, which leads me to believe they must contain a higher percentage of the pigmented chemical compounds.

---

1. Bechtold, Manian, Pham, *Handbook of Natural Colorants*, 2nd ed., 323.

Often confused as a polypore, *Echinodontium tinctorium* is a tooth fungus—if you look closely at a slice, you'll see the spines. This shelf-like conk, sometimes mistaken for a piece of wood, is a pathogen growing on hemlocks and firs that causes heart rot in those trees. Its common name, paint fungus, reflects its use by Chilkat, Syilx, Skwxwú7mesh, Nlaka'pamux, Kwakwaka'wakw, and other Indigenous peoples along the Northwest Coast of North America for aesthetic and symbolic body paint.[2] This mushroom prefers soaking in an alkaline (pH9) environment for weeks or months to achieve the deep, rich, dark burnt-red pigment and, as a dye, it will create a range of colors from salmon orange to pinkish red to purplish red.

---

2. Nancy J. Turner and Alain Cuerrier, "'Frog's Umbrella' and 'Ghost's Face Powder': The Cultural Roles of Mushrooms and Other Fungi for Canadian Indigenous Peoples," *Botany* 100, no. 2 (November 2021): https://doi.org/10.1139/cjb-2021-0052.

MUSHROOMS FEATURED IN THIS SECTION:

Echinodontium tinctorium

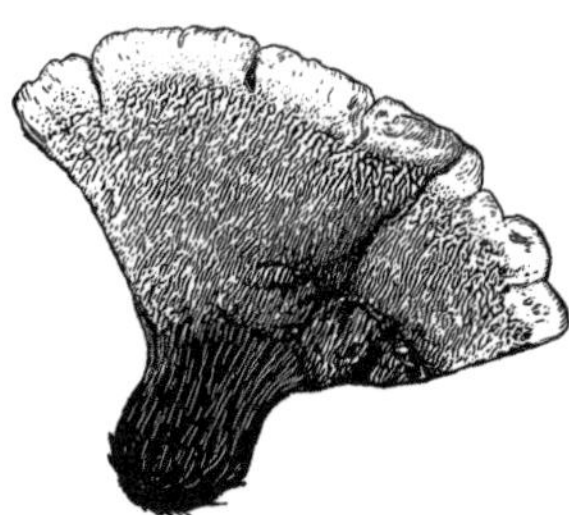

Hydnellum caeruleum

Hydnellum fuscoindicum

Hydnellum regium

Hydnellum suaveolens

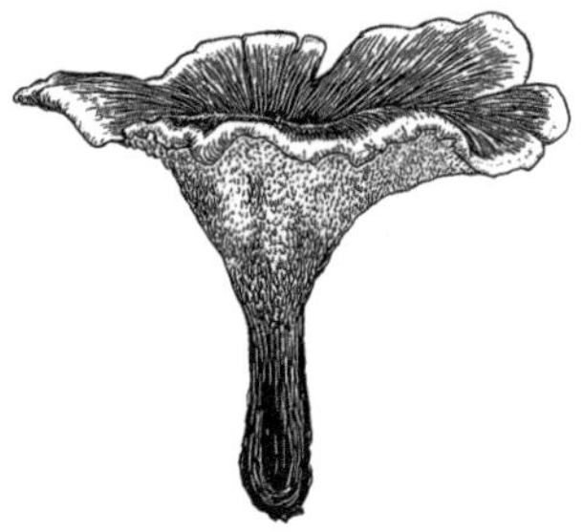

Phellodon niger

Sarcodon squamosus

Echinodontium tinctorium

GENUS: *Echinodontium*

SPECIES: *tinctorium*

COMMON NAME: Paint Fungus

Echinodontium tinctorium is a tough, woody, hoof-shaped mushroom with spiny teeth. Growing on the sides of downed conifer trees, this mushroom has a history of use by Chilkat, Syilx, Skwxwú7mesh, Nlaka'pamux, Kwakwaka'wakw, and other Indigenous peoples for its rusty red-brown pigment for aesthetic, symbolic, and medicinal purposes.[1] It is often mistaken as a polypore or a piece of wood. It requires a pre-soak in an alkaline (pH9) solution to coax out the rusty red-brown color.

HABITAT & ECOLOGY: Solitary or several growing under dead branch stubs on living conifer trees, specifically fir and hemlock. Often found fallen on the ground or on downed trees.

DISTRIBUTION: Western North America

SPORE PRINT: White

CAP: Olive-black, hoof shaped, with rough texture. Very tough, woody, bright orange to red-orange flesh.

HYMENIUM: Gray with blunt spines, often brittle and dry, with long darker tips.

STIPE: Lacks a stipe in a true sense; instead attaches to the trunks of trees.

---

1. Currier, Turner, "'Frog's Umbrella' and 'Ghost's Face Powder': The Cultural Roles of Mushrooms and Other Fungi for Canadian Indigenous Peoples," *Canadian Science Publishing*, 2022.

Dye Preparation

CONDITION	Dried
PART	Entire mushroom
PRE-SOAK	Yes
pH	9
RATIO	2:1
TEMP	160°F/71°C
TIME	1 hour

Pigment Preparation

CONDITION	Dried
PART	Entire mushroom
RINSE	Three times
BINDER	Gum Arabic
TEMP	85°F/29°C

PIGMENT

Base

Alum

Citric Acid

Iron

Copper Acetate

Soda Ash

Dye Preparation

CONDITION	Dried
PART	Entire mushroom
PRE-SOAK	Yes
pH	7
RATIO	2:1
TEMP	160°F/71°C
TIME	1.5 hours

Pigment Preparation

CONDITION	Dried
PART	Entire mushroom
RINSE	Three times
BINDER	Gum Arabic
TEMP	85°F/29°C

PIGMENT

Base

Alum

Citric Acid

Iron

Copper Acetate

Soda Ash

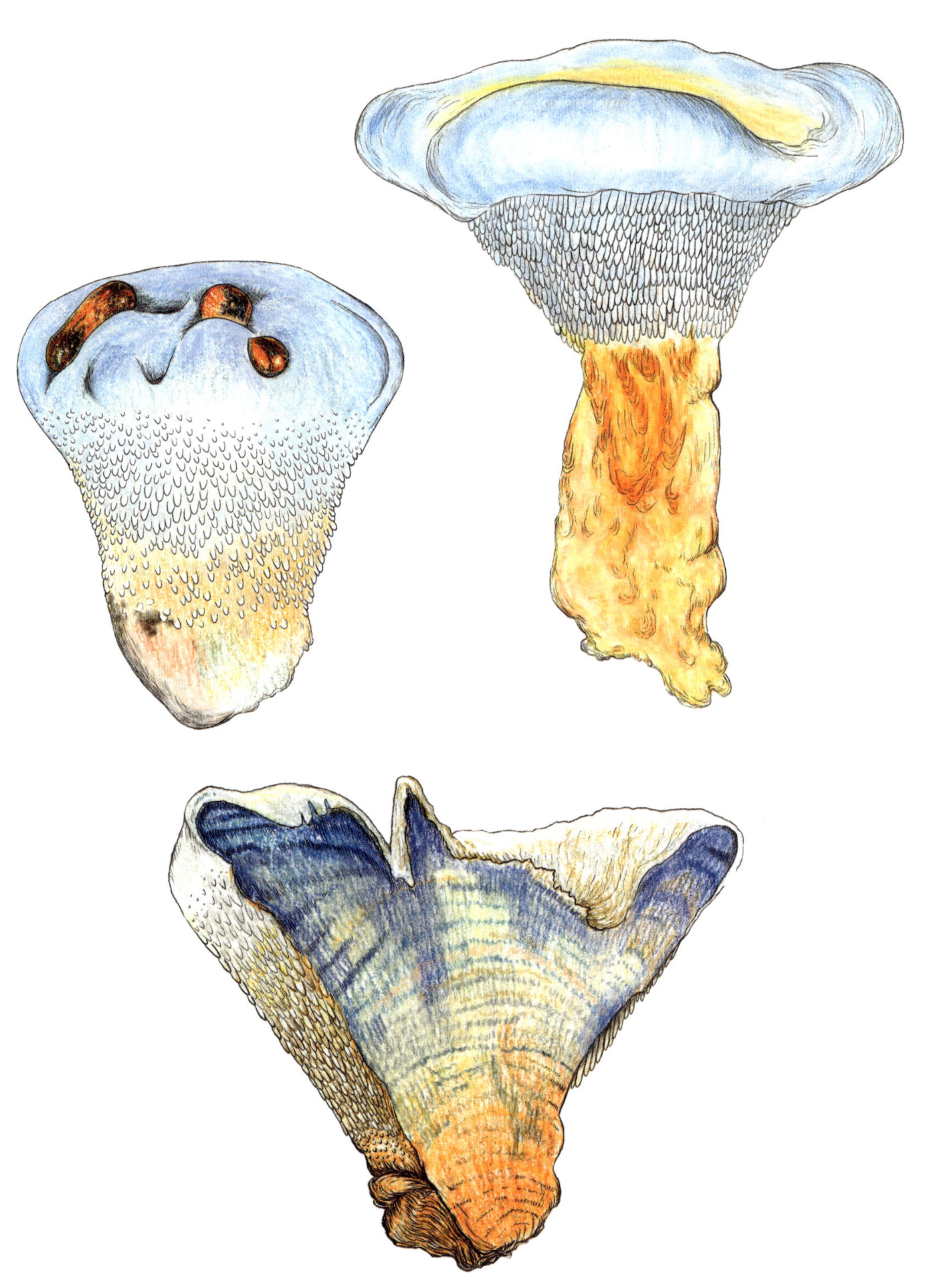

Hydnellum caeruleum

GENUS: *Hydnellum*

SPECIES: *caeruleum*

COMMON NAME: Blue-Orange Hydnellum

Hydnellum caeruleum is a tough, fibrous, light blue to dark blue mushroom with gray spines and an orange stipe. It grows on the ground in conifer forests. Cutting a cross section reveals blue flesh in the cap and orange in the stipe. It stains olive green to purple-black with KOH and requires an alkaline (pH9) solution to coax out the blue-green color.

HABITAT & ECOLOGY: Growing in clusters, solitary or scattered on the ground, under conifers and hardwoods.

DISTRIBUTION: Asia, Europe, North America

SPORE PRINT: Brown

CAP: Irregularly rounded and singular with sunken center. Often fused into larger clusters. Velvet surface, becoming ridged or pitted.

HYMENIUM: White to light gray spines, becoming brownish with age.

STIPE: Center to off-center, equal thickness throughout, slightly swollen at the base. Orange becomes darker with age.

Dye Preparation

CONDITION	Dried
PART	Entire mushroom
PRE-SOAK	–
pH	9
RATIO	2:1
TEMP	165°F/74°C
TIME	1.5 hours

Pigment Preparation

CONDITION	Dried
PART	Entire mushroom
RINSE	Twice
BINDER	Gum Arabic
TEMP	85°F/29°C

PIGMENT

Base

Alum

Citric Acid

Iron

Copper Acetate

Soda Ash

Hydnellum fuscoindicum

GENUS: *Hydnellum*

SPECIES: *fuscoindicum*

COMMON NAME: Violet Hedgehog

Hydnellum fuscoindicum is a purple-blue to blue-black mushroom. The pigment is contained throughout the entire mushroom. It has thick, firm flesh with a brittle cap and fibrous dark purple stipe and requires an alkaline (pH9) solution to coax out the blue-green color.

HABITAT & ECOLOGY: Solitary or scattered, growing on the ground under conifers and hardwoods.

DISTRIBUTION: Western North America

SPORE PRINT: Brown

CAP: A brittle cap that is purple-gray, becoming purple-blue to purple-black with age. Dry to moist surface, becoming rough and scaly in maturity.

HYMENIUM: Purple-gray spines become purple-blue with age. Spines are longer in the center, becoming short near the edges.

STIPE: Center to off-center, tapers to a pointed base. Deep purple-blue flesh, staining dark purple.

Dye Preparation

CONDITION	Fresh
PART	Entire mushroom
PRE-SOAK	Yes
pH	9
RATIO	2:1
TEMP	160°F/71°C
TIME	1.5 hours

Pigment Preparation

CONDITION	Fresh
PART	Entire mushroom
RINSE	Twice
BINDER	Gum Arabic
TEMP	85°F/29°C

PIGMENT

Base

Alum

Citric Acid

Iron

Copper Acetate

Soda Ash

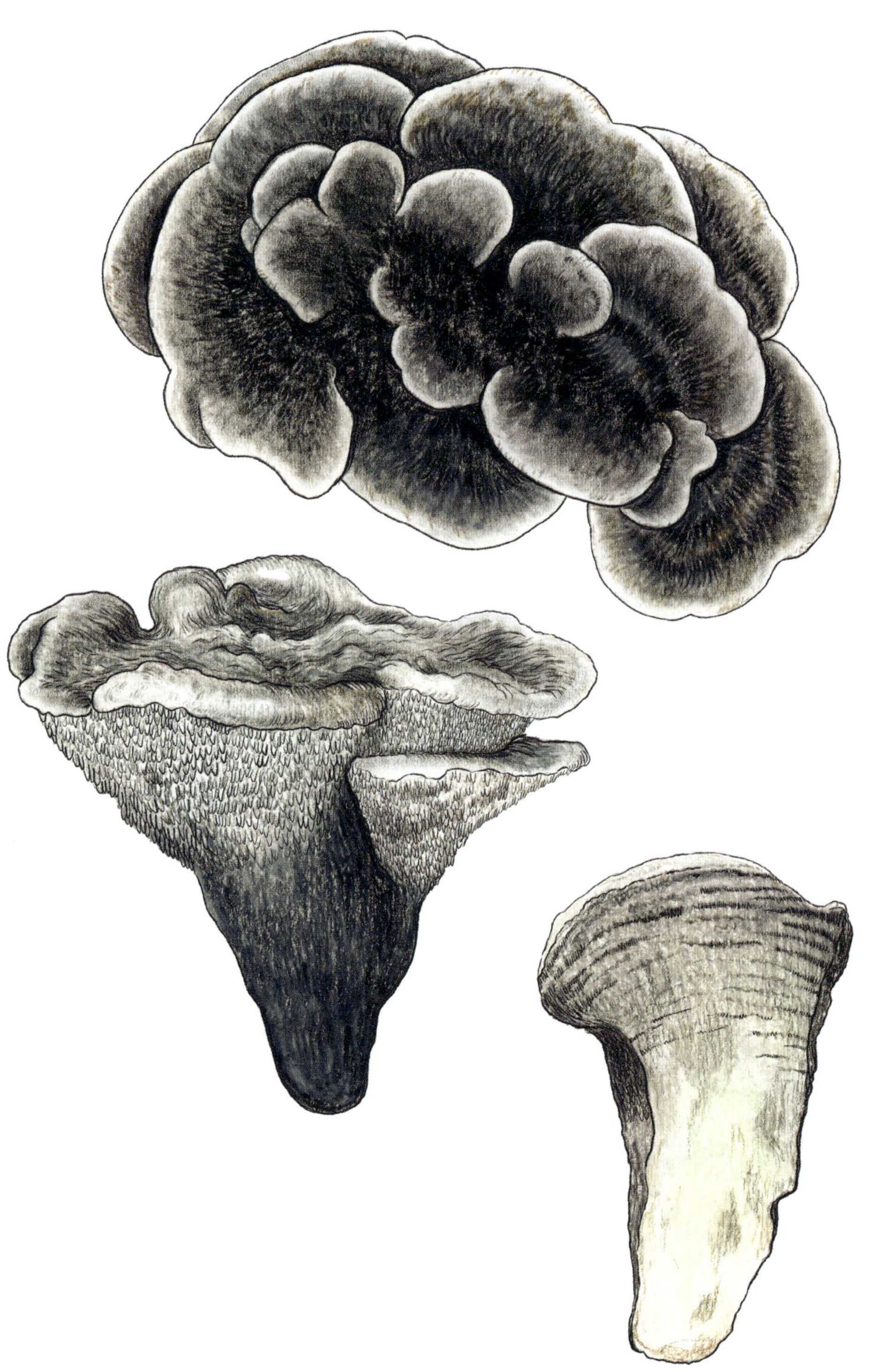

Hydnellum regium

GENUS: *Hydnellum*

SPECIES: *regium*

COMMON NAME: Bear Poop

Hydnellum regium is a dark black-violet mushroom. The pigment is contained throughout the entire mushroom. It has thick, firm flesh with small clusters of black-violet caps with a light gray rim, often fusing to form larger clusters. It requires an alkaline (pH9) solution to coax out the blue-green color.

HABITAT & ECOLOGY: Solitary or scattered, growing on the ground, under conifers, specifically spruce and pine.

DISTRIBUTION: Western North America

SPORE PRINT: Brown

CAP: A black-violet cap with a light gray rim. Often fused into larger clusters. Flesh is brown-gray with purple tones and brown-orange in the stipe.

HYMENIUM: Dark black-violet spines become darker with age.

STIPE: Center to off-center, equal thickness throughout, slightly swollen at the base. Deep brown-orange flesh.

Dye Preparation

CONDITION	Dried
PART	Entire mushroom
PRE-SOAK	–
pH	9
RATIO	1:1
TEMP	165°F/74°C
TIME	1 hour

Pigment Preparation

CONDITION	Dried
PART	Entire mushroom
RINSE	Three times
BINDER	Gum Arabic
TEMP	85°F/29°C

PIGMENT

Base

Alum

Citric Acid

Iron

Copper Acetate

Soda Ash

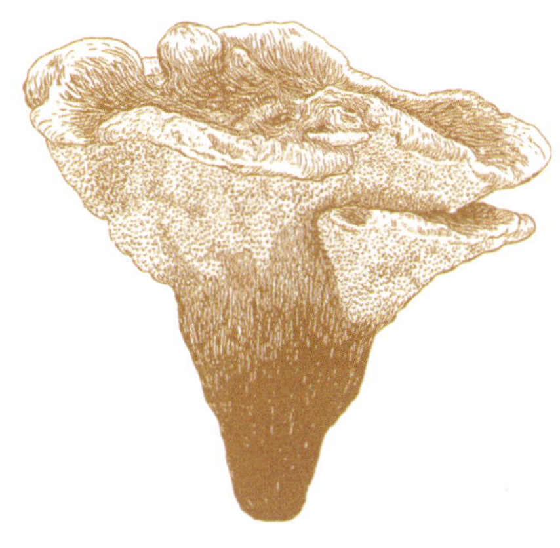

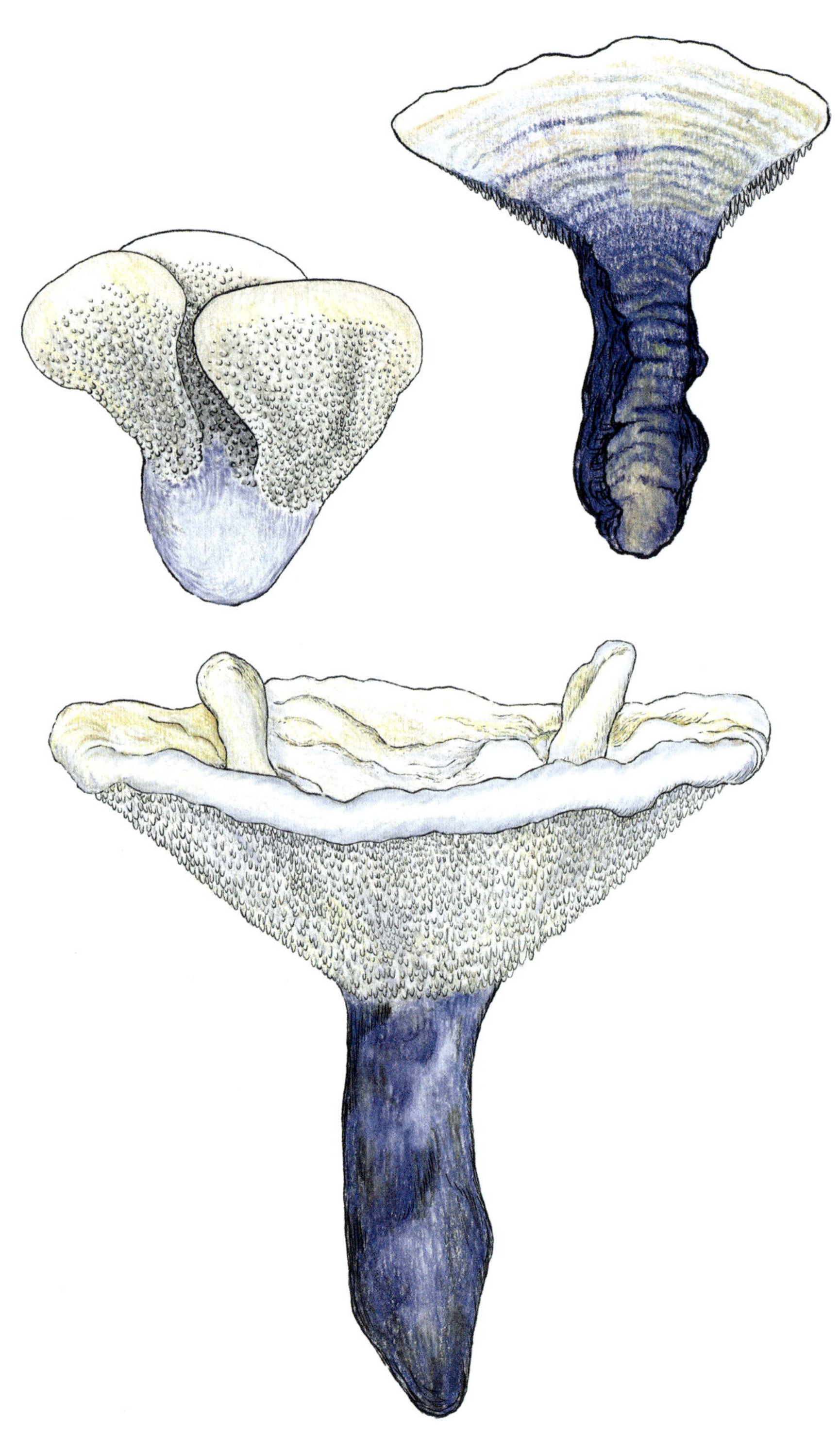

Hydnellum suaveolens

GENUS: *Hydnellum*

SPECIES: *suaveolens*

COMMON NAME:
Sweetgrass Hydnellum

Hydnellum suaveolens is a tough, fibrous white-gray to light white-brown mushroom with gray spines and a blue stipe. It grows in clusters on the ground in conifer forests. Cutting a cross section reveals light blue flesh in the cap and dark blue in the stipe. It requires an alkaline (pH9) solution to coax out the blue-green color and stains blue-green with KOH.

HABITAT & ECOLOGY: Growing in clusters, solitary or scattered on the ground in conifer forests, specifically spruce.

DISTRIBUTION: Asia, Europe, North America

SPORE PRINT: Brown

CAP: Irregularly rounded and singular with sunken center. Often fused into larger clusters. Velvet surface, becoming ridged or pitted.

HYMENIUM: White to light gray, becoming dark gray to gray-brown with age. Spines are longer in the center, becoming shorter near the edges.

STIPE: Center to off-center, equal thickness throughout, slightly swollen at the base. Light blue becomes dark blue to blue-black with age.

Dye Preparation

CONDITION	Dried
PART	Entire mushroom
PRE-SOAK	Yes
pH	9
RATIO	2:1
TEMP	165°F/74°C
TIME	1.5 hours

Pigment Preparation

CONDITION	Dried
PART	Entire mushroom
RINSE	Three times
BINDER	Gum Arabic
TEMP	85°F/29°C

PIGMENT

Base

Alum

Citric Acid

Iron

Copper Acetate

Soda Ash

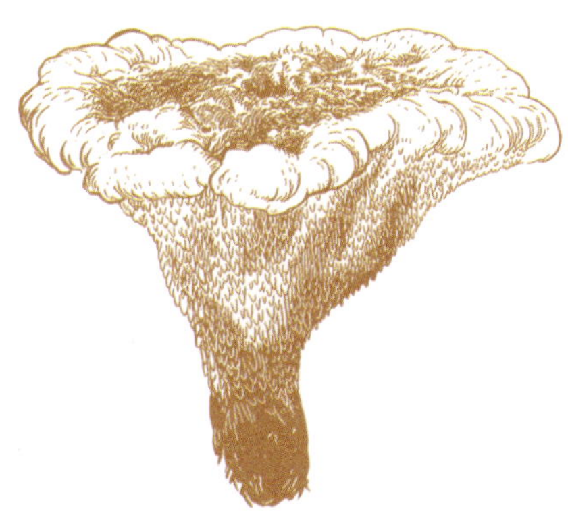

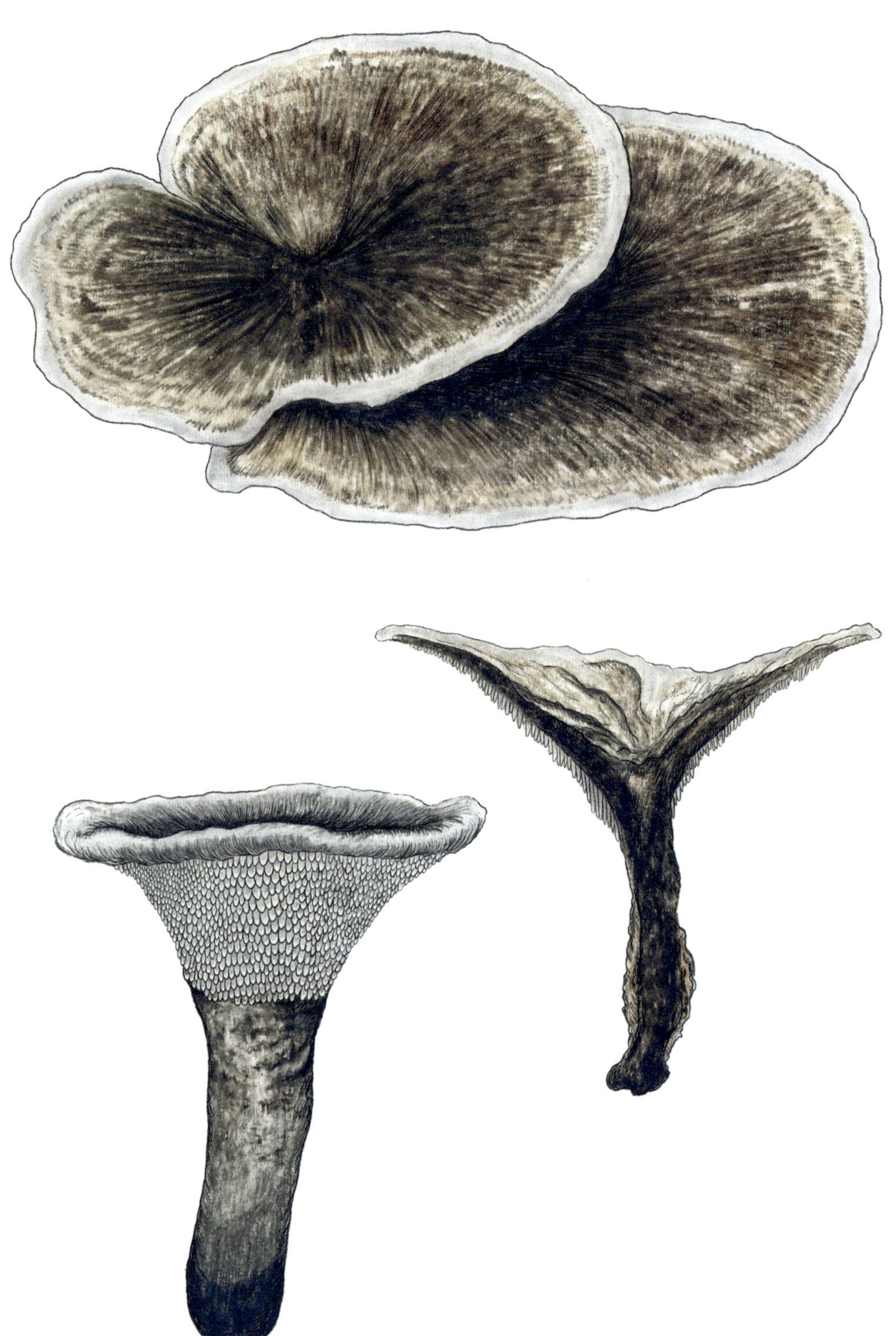

Phellodon niger

GENUS: *Phellodon*

SPECIES: *niger*

COMMON NAME: Black Tooth

Phellodon niger is a tough, fibrous mushroom with a velvety texture, becoming sunken in shape and wrinkled with ridges with age. It is light gray when young and becomes dark blue-black with age. It requires an alkaline (pH9) solution to coax out the blue-green color and stains blue-green with KOH.

HABITAT & ECOLOGY: Growing in clusters or solitary on the ground in conifer forests, specifically pine, hemlock, and spruce.

DISTRIBUTION: Europe, North America

SPORE PRINT: White

CAP: Wide and rounded, often fused into larger clusters, with a sunken center. Firm, tough, velvety. Light gray, becoming brown-gray and dark blue-black with age.

HYMENIUM: Light gray spines, darkening when bruising. Black flesh in the cap and stipe.

STIPE: Thick, centered, tapers down at the base, sometimes fused or branched, blue-black flesh.

Dye Preparation

CONDITION	Dried
PART	Entire mushroom
PRE-SOAK	–
pH	9
RATIO	1:1
TEMP	160°F/71°C
TIME	1.5 hours

Pigment Preparation

CONDITION	Dried
PART	Entire mushroom
RINSE	Three times
BINDER	Gum Arabic
TEMP	85°F/29°C

PIGMENT

Base

Alum

Citric Acid

Iron

Copper Acetate

Soda Ash

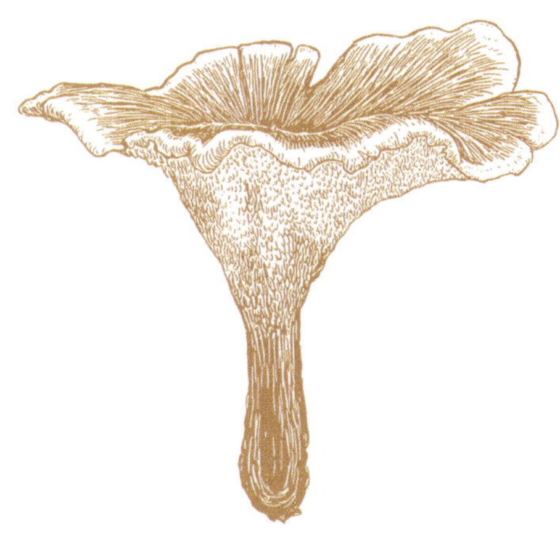

Sarcodon squamosus

GENUS: *Sarcodon*

SPECIES: *squamosus*

COMMON NAME:
Scaly Tooth

Sarcodon squamosus is dark brown to purplish brown-black, covered with tufted or shingle scales. *S. squamosus* is often confused with *Sarcodon imbricatus* because they are so similar. The non-dyer *S. imbricatus* grows primarily in association with spruce trees, and the dyer *S. squamosus* grows primarily in association with pine trees. It requires an alkaline (pH9) solution to coax out the blue-green color.

HABITAT & ECOLOGY: Solitary or scattered in groups, sometimes forming an arc on the ground. Growing in conifer forests, specifically pine.

DISTRIBUTION: Europe, North America

SPORE PRINT: Brown

CAP: Irregularly rounded at first with sunken center, becoming slightly uplifted with age. Brown to purplish brown-black with buff to light brown color between scales. Surface is covered with tufted scales, dry to moist, and often very large in size.

HYMENIUM: Grayish to light brown spines become all brown with age. Spines are longer in the center, becoming short near the edges.

STIPE: Center to off-center, equal thickness throughout, slightly swollen at the base. Light brown, turning dark brown at the base.

Dye Preparation

CONDITION	Dried
PART	Entire mushroom
PRE-SOAK	–
pH	9.5
RATIO	2:1
TEMP	160°F/71°C
TIME	1 hour

Pigment Preparation

CONDITION	Dried
PART	Entire mushroom
RINSE	Three times
BINDER	Gum Arabic
TEMP	85°F/29°C

PIGMENT

Base

Alum

Citric Acid

Iron

Copper Acetate

Soda Ash

CHAPTER FIVE

Odds & Ends

This is always my favorite section of any mushroom field guide because it is filled with unique, curious, and fascinating specimens. This book would not be complete without featuring these singular representative mushrooms from each category type: False Gilled, Puffball, Coral, and Conk. One of my absolute favorite mushrooms is a false-gilled mushroom (meaning the spore-bearing surface is on ridges rather than gills). This mushroom is commonly known as a blue chanterelle, though it is not related to the *Cantharellus* genus. Rather, it is in the *Polyozellus* genus, specifically *P. atrolazulinus*. The blue chanterelle is a great example of "what you see is what you get"—the color on the outside of the mushroom is the same as the extracted colorant of the chemical pigment compound. The mushroom itself is a deep, dark blue with hues of light gray-blue around the rim of each florette. It produces an array of green-blue hues both as a dye and a pigment. Plus, it is an edible mushroom (although not one of my favorites; I much prefer it for its colors). The blue chanterelle's beautiful, tightly packed clusters are reminiscent of flower florets with long fluted stalks. I wish finding them was as easy as popping into the florist's shop!

Puffballs are common mushrooms you may have come across in open fields or even in your front yard. They do not disperse spores from beneath their caps, but rather have an internal spore mass that becomes powdery as it ages. When a mature puffball's outer skin is disturbed by wind or drops of rain, these mushrooms release a visible poof of spores into the air. The well-known exceptional dye mushroom *Pisolithus arhizus* is a deep brown puffball. When you find these mushrooms at the midpoint in their growth cycle, they are squishy, gooey masses, and slicing them open reveals a flush of colors: yellow, orange, and green. These colorants belong to the badione chemical compound family and are responsible for striking golden and red-brown hues. Working with the hydrophobic spores of *P. arhizus* can be challenging. I have found emulsifying them for a day or two in a sealable plastic bag with water and a drop of soap will help hydrate the powdery spores, preparing them for the dye bath.

If you ever find a coral mushroom, you will see the striking resemblance to its namesake, the marine invertebrates that grow in reefs in our oceans. These colorful fungi develop upright branches from a fleshy base, or trunk. It is easy to visually identify most members of this genus—called *Ramaria*—as belonging to the coral mushroom group; however, identifying particular species with 100% certainty can be incredibly challenging. As a rule of thumb, the more colorful orange and yellow-coral mushrooms will produce a beautiful violet-gray in conjunction with iron mordant. However, the trick is to work with these mushrooms fresh or frozen; dried coral mushrooms will not create much more than a dull, light beige color.

Inonotus obliquus is commonly known as chaga, a traditionally revered medicinal in Asia and Eastern Europe that is also receiving attention from health-conscious folks in the West. Technically not a mushroom, but rather a canker, it takes on a conk-type shape and can be found growing on birch trees throughout the Northern Hemisphere. Chaga owes its recent popularity to its medicinal properties—clinical studies suggest it has anti-tumor, antioxidant, and antiviral potential. It may soon be inoculated into forest land at scale to generate new income opportunities for forest owners by producing nutritional supplements.[1] *I. obliquus* is a very hard black canker that is golden reddish brown on the inside. It is a variety of white rot fungus, so it most likely contains cinnamic acid as well as other substances such as hispidin and hypholomin, which contain yellow and orange pigment compounds. Giving it an acidic (pH4) dye bath solution helps it produce brilliant yellows on silk fiber.

MUSHROOMS FEATURED IN THIS SECTION:

Inonotus obliquus

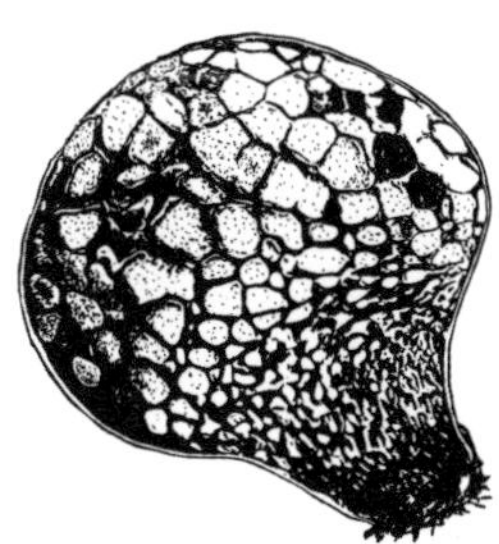

Pisolithus arhizus

Polyozellus atrolazulinus

Ramaria spp.

---

1. Jari Miina et al. "Inoculation Success of *Inonotus obliquus* in Living Birch (*Betula* spp.)," *Forest Ecology and Management* 492 (July 2021): https://www.sciencedirect.com/science/article/pii/S0378112721003327.

Inonotus obliquus

GENUS: *Inonotus*

SPECIES: *obliquus*

COMMON NAME: Chaga

Inonotus obliquus is technically not a mushroom, but rather a blackened, cracked canker parasite growing on trees. It has a dark red to brown-black exterior and rusty brown to yellow-orange interior.

HABITAT & ECOLOGY: Growing on well-developed hardwood trees, most commonly birch, sometimes beech and elm. Very slow to mature.

DISTRIBUTION: Asia, Europe, North America

SPORE PRINT: Not applicable

CAP: Black, gnarly with a rough texture, growing out of wounds in the trunks of trees. Rusty brown to yellow-orange interior flesh.

HYMENIUM: Does not have a spore-bearing surface.

STIPE: Does not have a stipe. It is an irregular-shaped canker growing on the trunks of trees.

Dye Preparation

CONDITION	Dried
PART	Entire mushroom
PRE-SOAK	–
pH	5
RATIO	1:1
TEMP	165°F/74°C
TIME	1 hour

Pigment Preparation

CONDITION	Dried
PART	Entire mushroom
RINSE	Twice
BINDER	Gum Arabic
TEMP	85°F/29°C

PIGMENT

Base

Alum

Citric Acid

Iron

Copper Acetate

Soda Ash

Pisolithus arhizus

GENUS: *Pisolithus*

SPECIES: *arhizus*

COMMON NAME: Dyer's Puffball

Pisolithus arhizus is easily recognizable and lovingly referred to as dog turd or dead man's foot. Cutting a cross section reveals bright yellow, orange-brown and olive-black pigmentation. With age it becomes a rusty red-brown powdery mass.

HABITAT & ECOLOGY: Widely distributed, growing in nutrient-deprived soils, disturbed areas, and open woods. Often found under a variety of trees including birch, oak, pine, and others.

DISTRIBUTION: North America

SPORE PRINT: Rusty red-brown

CAP: Orange-brown to brown, club-like shape becoming quite irregular with age.

HYMENIUM: Small oblong spore packets range in color from bright yellow, orange-brown to olive-black. Disintegrates into a brown powdery mass of spores.

STIPE: Thick, stout, deep rooting, stalklike base.

Dye Preparation

CONDITION	Dried
PART	Entire mushroom
PRE-SOAK	Yes
pH	7
RATIO	1:1
TEMP	160°F/71°C
TIME	1.5 hours

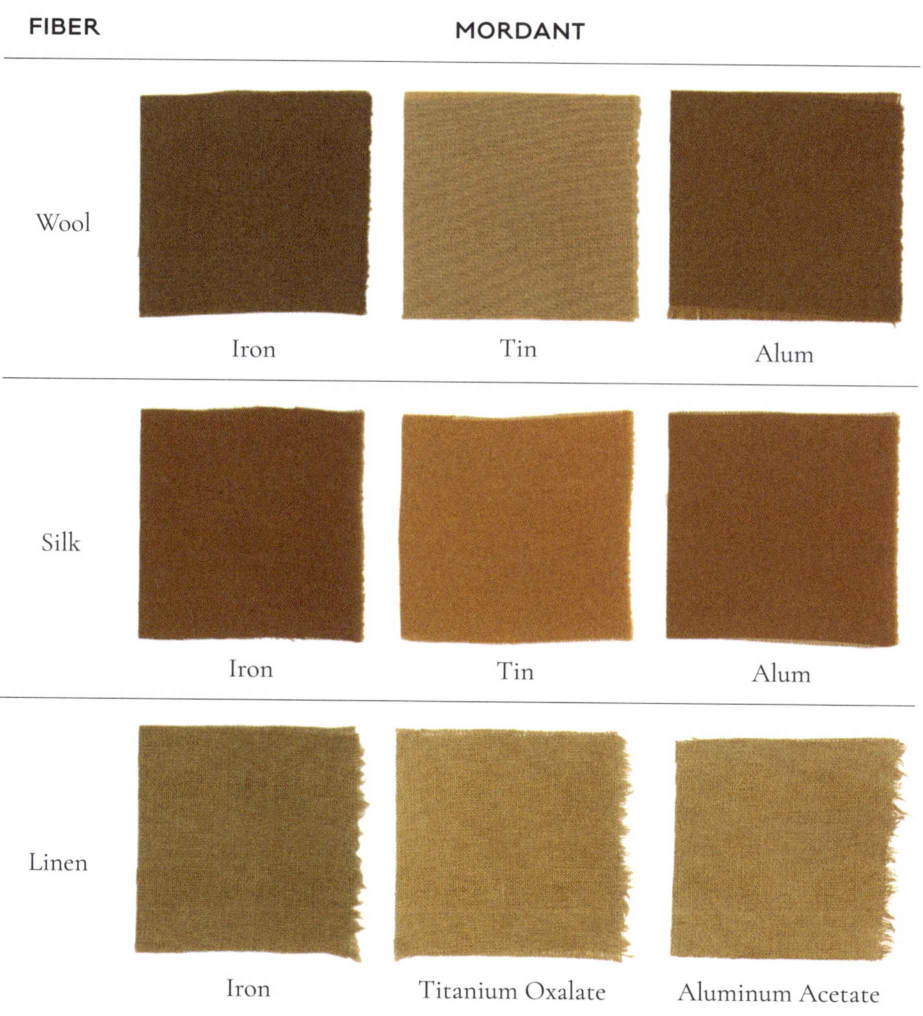

Pigment Preparation

CONDITION	Dried
PART	Entire mushroom
RINSE	Once
BINDER	Gum Arabic
TEMP	85°F/29°C

PIGMENT

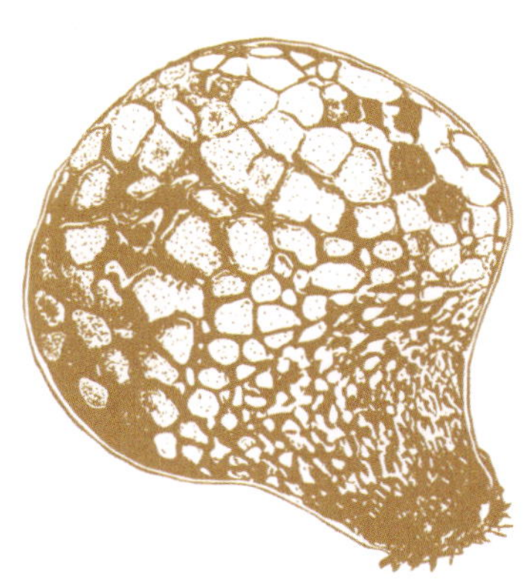

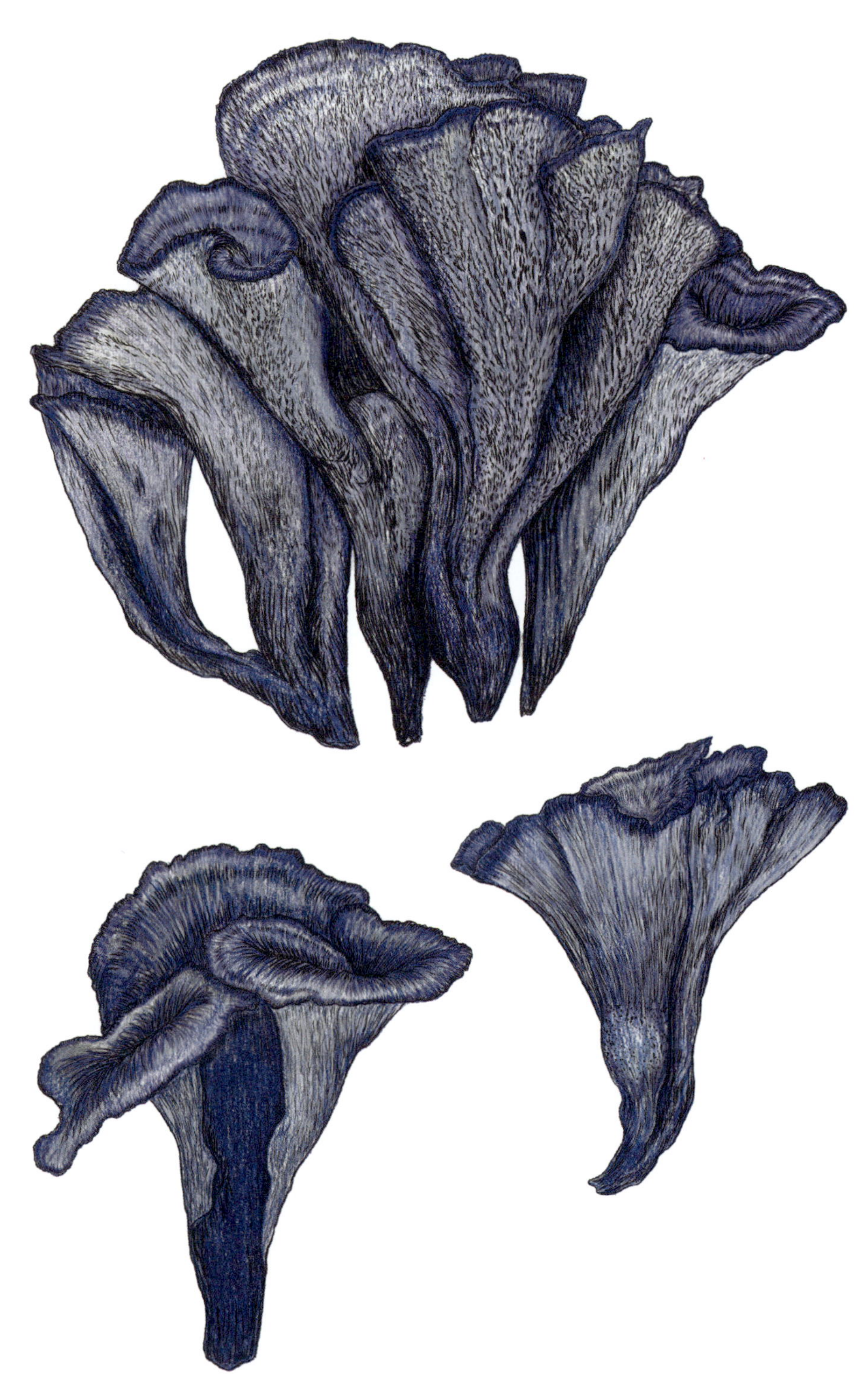

Polyozellus atrolazulinus

GENUS: *Polyozellus*

SPECIES: *atrolazulinus*

COMMON NAME: Blue Chanterelle

Polyozellus atrolazulinus is deep purple-blue to blue-black with a light gray-blue rim. Growing on the ground in clumps with fluted-shaped stipes and spoon-shaped caps, it has tough flesh with gilled ridges down the stipe and stains dark olive-black with KOH.

HABITAT & ECOLOGY: Grows in clumps on the ground in conifer forests, specifically spruce, fir, and hemlock.

DISTRIBUTION: North America

SPORE PRINT: White

CAP: Deep purple-blue to blue-black spoon-shaped caps. Smooth surface, moist to dry, with wrinkles around the edge.

HYMENIUM: Deep purple-blue to blue-black with gilled ridges running down the stipe into a fused base.

STIPE: Thick, short, usually fused at the base, deep purple-blue to blue-black.

Dye Preparation

CONDITION	Dried
PART	Entire mushroom
PRE-SOAK	–
pH	7
RATIO	1:1
TEMP	165°F/74°C
TIME	1 hour

FIBER	MORDANT		
Wool	Iron	Tin	Alum
Silk	Iron	Tin	Alum
Linen	Iron	Titanium Oxalate	Aluminum Acetate

Pigment Preparation

CONDITION	Dried
PART	Entire mushroom
RINSE	Twice
BINDER	Gum Arabic
TEMP	85°F/29°C

PIGMENT

Base

Alum

Citric Acid

Iron

Copper Acetate

Soda Ash

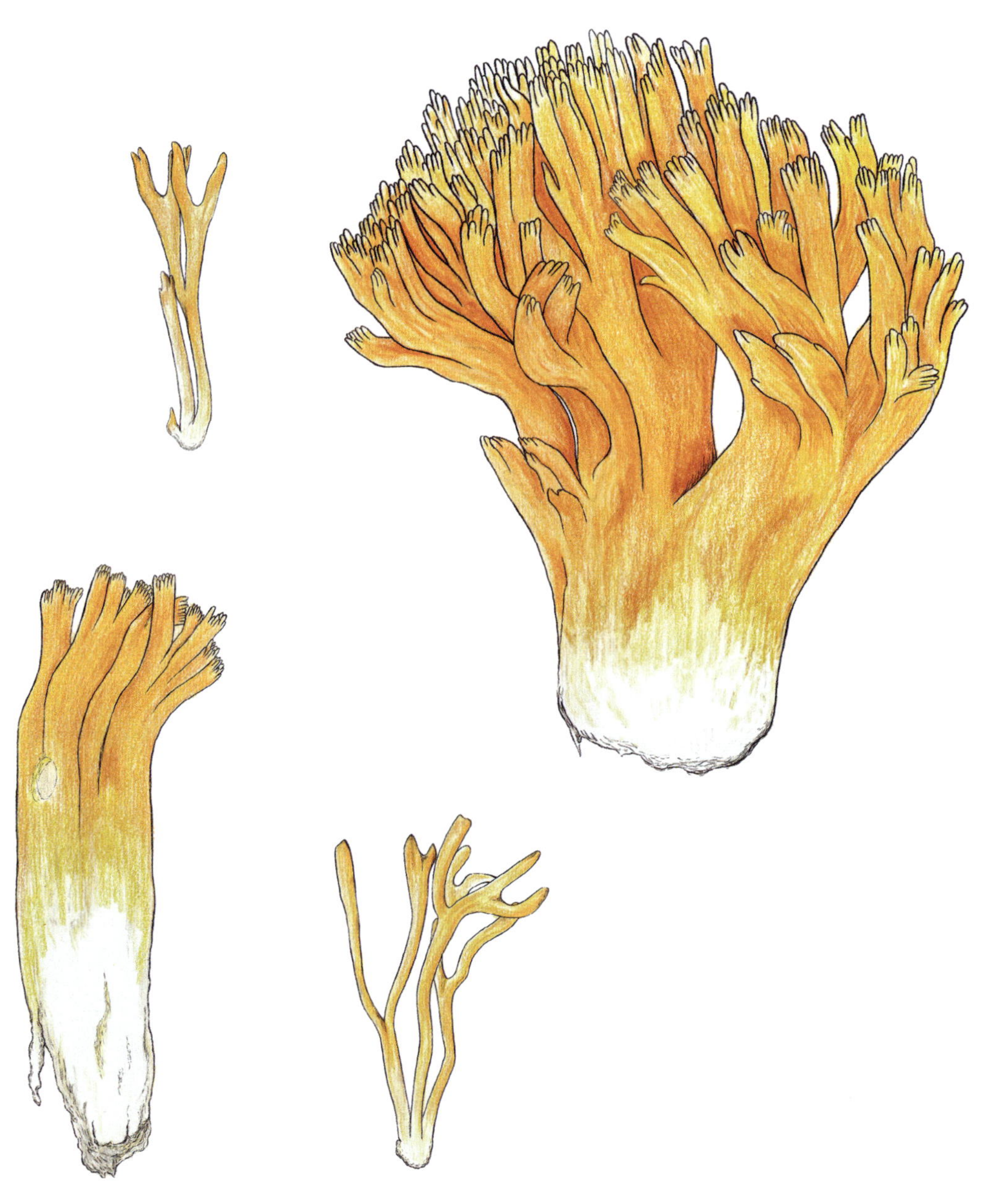

Ramaria species

GENUS: *Ramaria*

SPECIES: All species

COMMON NAME: Coral

Commonly known as coral mushrooms, it is very difficult to identify species of the *Ramaria* genus without working under a microscope. The spines of these mushrooms point upward with diverse branching patterns and colorations. Whether fresh or frozen, colorful *Ramaria* in combination with iron creates a smoky purple-gray color.

HABITAT & ECOLOGY: Often solitary, scattered, or in masses, growing in conifer forests.

DISTRIBUTION: Asia, Australia, Europe, North Africa, North America

SPORE PRINT: Ranging from orange to yellow to red-brown.

CAP: Ranging from slender upright branches to multiple branches arising from a common base. Coloration ranging from white to beige, yellow to orange.

HYMENIUM: Flesh can be soft, firm, thick, or brittle and ranges in size and color.

STIPE: Often short, thick, fused, and tapering downward.

Dye Preparation

CONDITION	Fresh
PART	Entire mushroom
PRE-SOAK	–
pH	7
RATIO	1:1
TEMP	160°F/71°C
TIME	1.5 hours

Pigment Preparation

CONDITION	Fresh
PART	Entire mushroom
RINSE	Twice
BINDER	Gum Arabic
TEMP	85°F/29°C

PIGMENT

Base

Alum

Citric Acid

Iron

Copper Acetate

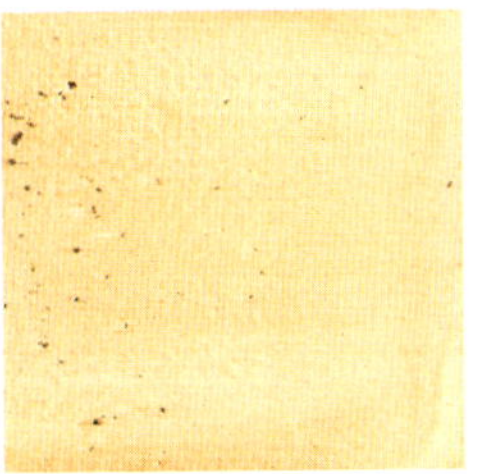
Soda Ash

SECTION III

PROCESS

The process of making artists' materials from mushrooms helps me connect with wild and natural places. Foraging means slowing down and paying attention to nature's subtle cues. I am learning the rhythms and intricacies of the forests near my home, and I feel inspired to practice careful stewardship of the lands I inhabit. Since my own artworks are born of materials I've carefully harvested, I like to think they are imbued with some of nature's beauty and richness. I marvel at the array of colors that can be derived from the natural world—it is a gift to be able to use them for my own creative expression.

MAKING COLOR

Transforming mushrooms into colors is relatively simple. You can process them in a dye bath, just like you steep tea leaves in a cup of warm water. You can transform your dye bath into a pigment by adding a mordant and precipitant to make a lake pigment, which is the result of a liquid, soluble dye bath turning into a solid, insoluble pigment. You can whirl the powdered lake pigment into paint or ink by mulling it in a circular motion on a slab until it is coated with the binder. If you are just starting your journey into making natural artists' colors, you will need a few basic pieces of equipment. You do not need to spend a lot of money to get started; almost all of the supplies can be found at your local thrift shop.

DYE EQUIPMENT

Bowl
Cooktop
Food scale
Measuring spoons
Mordants
pH strips
Pot
Soap
Stir stick
Strainer
Thermometer

LAKE-MAKING EQUIPMENT

Bowl
Cooktop
Food scale
Glass jars
Measuring spoons
Mordant and precipitant
Stir stick

PAINT-MAKING EQUIPMENT

Binder
Measuring spoons
Mortar and pestle
Muller
Palette knife
Pipette dropper
Slab

Best practices are that these tools should be dedicated to this non-food purpose, not shared with your kitchen tools. You don't need much space to set up your working area, but you do need a cooking surface and good ventilation (some mushrooms stink!). Working outdoors, if you have the option, isn't a bad idea.

The supporting photographs for each process section illustrate the few basic items you can use to coax colors from mushrooms. These photographs were taken in an indoor studio setting, so my outdoor stove is not included.

A JOURNEY OF DISCOVERY

This book offers formulas that produce repeatable results that you can use as a starting point to create your own recipes. I encourage you to experiment with extracting colors. There is no universal recipe or rule for how best to get color from mushrooms, and sometimes the process feels a little like magic, but paying attention to these specific factors can help ensure consistent, repeatable results:

- Fiber pre-treatment and mordants
- Condition of the mushroom
- Mushroom parts used
- Ratio of weight of goods to weight of fiber (WOG:WOF)
- Length of time cooking
- Temperature of dye bath
- pH of dye bath
- Finishing of the fiber

ON
OFF
Mode
5.0
g
Pcs
Tare

CHAPTER ONE

Preparation

WORKING WITH MUSHROOMS

The best way to collect mushrooms that are filled with pigment is to forage for them yourself, though you may be able to find wild mushrooms that are both edible and good for dyeing at local farmers' markets. Once you have acquired some pigment-bearing mushrooms, the process of coaxing color out of them is particular to each genus and species of mushroom. You can work with mushrooms in various forms to create color: They can be fresh, dried, powdered, frozen, freeze-dried, or pre-soaked in an alkaline (pH9) bath or in an acidic (pH4) bath. Many mushrooms hold multiple pigments. If pigments are not isolated, the different colorants will combine to create unique colors. This book focuses on getting colors from mushrooms in the form of dyes and lake pigments, but there are many other natural artists' materials that you can make from fungi, including paper and crayons. I also use mushrooms for bundle steam dyeing and botanical printing. Many more uses are awaiting your exploration. Have fun!

WEIGHING MUSHROOMS

When working with mushrooms for dye, everything comes down to weight. The proportion of mushrooms to fiber determines the hue and saturation of the dye. Too few mushrooms and too much fiber means you won't get much color. Here is the rule of thumb for natural dyes, where mushrooms are the dye goods: The weight of the goods (WOG) should equal the weight of fiber (WOF). WOG = WOF as a formula; or, as a ratio, 1:1. There are, of course, exceptions to this rule. Some mushroom species yield a disproportionate amount of color, while some produce only scant color. If you have a mushroom that is a strong dyer, such *Phaeolus schweinitzii*, you will only need half as much to yield a good dye result, a ratio of 1:2. If you have a mushroom that is a weaker dyer, such as *Hypomyces lactifluorum*, you may need to double its weight in relation to the fiber, a ratio of 2:1.

Fresh and frozen mushrooms work well for dyeing, and (when doing a side-by-side comparison of dried vs. fresh) some mushrooms, such as *Hypomyces lactifluorum*, yield deeper, richer colors when fresh. In fact, certain species, such as *Ramaria*, don't produce any dye color if they have been dried but are good dyers when fresh or frozen. Accurately weighing fresh mushrooms presents its own complexities: Water accounts for the majority of the weight of fresh or frozen mushrooms and varies in quantity depending on the condition of the specimens. For this reason, many people choose to work with dried mushrooms as a way of standardizing weights. Mushrooms can be dried in a (dedicated!) dehydrator at a low temperature for an extended period of time. Be sure to keep an eye on the temperature. If it gets too hot, you can overcook the mushroom and brown out the pigment. If you do use fresh or frozen mushrooms, one trick is to eyeball your pile of mushrooms, and if it is the same size or larger than your wadded-up pile of fiber, you should have enough mushrooms to get color.

PREPARING MUSHROOMS

To make a potent dye bath, it is important to get as much of the mushroom's surface area exposed to the water as possible. Grinding dried mushrooms into powder with a mortar and pestle or a (dedicated!) grinder certainly increases surface area but isn't really necessary. It also works well to chop mushrooms into very small pieces with a sharp knife before dehydrating them or using them fresh. Some mushrooms prefer to be soaked in water or a modified alkaline (pH9) or acidic (pH5) solution for a few days or weeks to help break down their fibrous structures and release more pigment.

Take appropriate safety precautions when preparing specimens for the dye bath. It is perfectly safe to handle all kinds of mushrooms, but never ingest any part of a mushroom unless you are 100% confident in your identification; some are quite toxic and poisonous. When working with any powdered material (including dust from mushrooms), it is best practice to protect your lungs, eyes, and skin, so wear a good quality mask, protective eyewear, and gloves.

CHAPTER TWO

Dyes

CREATING DYES

Mushroom dyes fall into different categories, the two main ones being *adjective* and *substantive*. The majority of mushrooms produce adjective dyes, meaning it is necessary to mordant the fibers in order for the dyes to properly take. The term *mordant* comes from the Latin *mordere*, which means "to bite," and is the process of pre-treating fiber before dyeing to ensure the colorant will bind to—or "bite"—the fiber. Adding a mordant not only helps set the dye but also improves lightfastness and allows us to pull a broad range of hues from each species of mushroom. Substantive dyes can be fixed within fibers without the addition of any other substances, like mordants, making them the simplest to work with. Keep in mind that you won't be able to achieve the same range of colors without a mordant, and with some mushrooms, the substantive dye colors may subtly shift. While it is true that some natural dyes have inferior lightfastness compared to their synthetic counterparts, meaning they tend to fade or change color with prolonged exposure to light, mushroom dyes have shown promise in their light resilience.[1]

1. Rice, *Mushrooms for Color*, iv.

PREPARING FIBERS

The material a fiber is made with, as well as its weave, density, color, and weight, will determine how well it receives the dye. Starting with clean, scoured (washed) fibers that are prepared to accept dye is a critical step in the process. There are several ways to scour fibers, but these are my preferred methods. Depending on the fibers being used, it is best to scour them with a neutral soap (pH7) at a temperature that does not damage the fibers.

- **WOOL:** Orvus paste at 1% WOF at 160°F/71°C for 1 hour.
- **SILK:** Orvus paste at 1% WOF at 140°F/60°C for 30 minutes.
- **CELLULOSE:** Synthrapol soap at 1% WOF + soda ash at 1% WOF at 180°F/82°C for 1 hour.

MORDANTING FIBERS

Most natural colorants need a mordant to be added to the fiber. This process of pre-treating fiber with a mordant before dyeing will help ensure the colorant will bind to—or "bite"—the fiber. The type of mordant used will often modify the color, resulting in a range of hues from one mushroom species across different fiber types. The details below are my preferred methods; however, these are not the only ways to mordant fibers.

- **WOOL:** Water warmed, mordant dissolved, fibers dropped in, and everything heated up to 160°F/71°C. Pull the pot off the burner and let it cool with the fibers inside. Stir gently and often. Leave everything in the pot for 24 hours, after which remove the fibers and rinse.

- **SILK:** Water warmed, mordant dissolved, fibers dropped in, and everything heated up to 140°F/60°C. Pull the pot off the burner and let it cool with the fibers. Stir gently and often. Leave everything in the pot for 24 hours, after which remove the fibers and rinse.

- **CELLULOSE:** Water warmed, mordant dissolved, fibers dropped in and everything heated up to 180°F/82°C. Pull the pot off the burner and let it cool with the fibers. Stir gently and often. Leave everything in the pot for 24 hours, after which remove the fibers and rinse.

- **CELLULOSE WITH ALUMINUM ACETATE:** Working with cellulose fibers requires a two-step process to get aluminum acetate to bind with the fiber; however, this process is not needed for other types of mordants with cellulose fibers. I prefer working with a tannin bath (specifically oak gall nut) to affix the aluminum acetate in a mushroom dye bath.

Keep the water for the tannin bath below 100°F/38°C (any warmer and the tannin will darken, coloring the fiber). Once the tannin is dissolved and fibers have been dropped in, stir gently and often, and soak for 2 hours, then remove the fiber and shake or wick away excess water. *Do not rinse*. Immediately put the wet fibers into a pot of warm tap water with dissolved aluminum acetate. Stir gently and often. Leave everything in the pot to cool for 24 hours, after which remove the fibers and rinse.

- **IRON:** A note about working with iron: It is a very strong contaminant and can easily stain other fibers. When using iron as a mordant, take caution not to contaminate other mordanted fibers or dye baths. If planning to combine iron-mordanted fibers with other mordanted fibers in the same dye pot, best practices are to put the well-rinsed iron-mordanted fibers into a pot of clean water, bring it up to the appropriate temperature for the fiber type, and maintain that temperature for 10 minutes. This process allows any residual iron to either bond with the fiber or be left in the water. Remove the fibers, rinse well, and drop them into the dye pot. Please note iron in the form of ferrous sulfate is not safe for babies, small children, or pets so please keep it out of reach.

A variety of different types of mordants can be used successfully; however, some mordants, such as chrome and copper, are toxic to humans and aquatic life. Mushrooms themselves can even be used as mordants. This process dates back to the fifteenth century and includes using *Laricifomes officinalis* (agarikon), *Fomes fomentarius* (hoof fungus), and *Polyporus alveolaris* (mulberry polypore) to set dyes in cloth.[2] Below are my preferred methods, but these are not the only mordants that can be used.

---

2. Cardon, *Natural Dyes*, 525.

- **ALUMINUM POTASSIUM SULFATE:** (Alum) 10% WOF
- **ALUMINUM ACETATE:** 6% WOF for cellulose fiber only
- **FERROUS SULFATE:** (Iron) 2% WOF
- **STANNOUS CHLORIDE:** (Tin) 3% WOF
- **TITANIUM OXALATE:** 6% WOF for cellulose fiber only
- **TANNIN:** (Gall Nut Extract) 10% WOF for cellulose fiber in combination with aluminum acetate. Use approximately a 30:1 (water:fiber) ratio for the tannin bath in order to balance the amount of tannin on the fiber and in the water.[3]

---

3. Joy Boutrup, Catharine Ellis, *The Art and Science of Natural Dyes: Principles, Experiments, and Results* (Atglen: Schiffer Craft, 2019), 127.

MAKING DYE BATHS

Creating a mushroom dye bath is like steeping tea leaves in a cup of warm water. Place the prepared mushrooms into a pot, add water, slowly bring up to temperature on a cooktop, let cook for at least an hour, strain out mushroom bits, add wetted fibers to the dye bath and continue cooking for about an hour, remove, rinse, and hang to dry. Voila! The details below are my methods; however, there are many different processes for dyeing fiber.

1. **WATER:** Water quality is a key variable that can inform and potentially modify the colors of the dye. Well water that is heavy in iron creates darkened, saddened colors or can make different colors altogether. Hard water rich in calcium can affect the subtle chemistry of the dye bath and alter colors. Working with captured rainwater or distilled water creates a more consistent baseline.

2. **FIBERS:** Soak prepared fibers in water before putting them in the dye bath to achieve a more even color distribution. It might take some time to thoroughly wet the fiber, from several hours up to overnight. Gently heating the water will help speed up the process.

3. **POT:** Work with any inexpensive stainless steel pot and, for smaller dye baths, consider using glass canning jars. Fill the vessel with only enough water to cover the fibers and allow them to freely move around the dye pot. (It's important to be environmentally conscientious about how much water you're using.)

4. **MUSHROOMS:** It is always a good idea to put chopped or ground mushrooms in a strainer bag or to strain out mushroom solids from the dye bath before adding the fiber. This helps ensure the fiber stays as clean as possible during the dyeing process. If the strained mushrooms are not depleted of their color, they can be frozen and saved for future use in a new dye bath.

5. **TEMPERATURE:** Being precise with temperatures is one of the most important factors for color success. With some mushrooms, if they get too hot or stay too hot for too long, the color can be zapped or browned out. But some species need to get up to a certain temperature in order to release the color. Keeping an eye on the temperature and maintaining the dye bath between 140–165°F/60–74°C during the cooking time will ensure consistent results and great color.

6. **COOKING:** With most mushrooms, cooking them for about an hour will result in a potent dye bath; however, some tougher mushrooms need a little longer cooking time to break down and release their colors. Use your eyes and your intuition.

7. **DYEING:** Once the fibers are in the dye pot, they typically need about an hour to get a good saturation of color to bind to the fiber. However, there are plenty of exceptions to this rule of thumb. Sometimes the saturation of color is achieved in 15 to 30 minutes, but the combination of heat, time, dye, and fiber may not have had enough time to produce a bond. If this happens, keep cooking the fibers in a clear pot of water at the same temperature until it reaches at least one hour of cooking time in total.

8. **STORAGE:** Keep in mind that the dye bath does not need to be used immediately; it can be stored in an airtight, lightproof container and used at another time. If you are going to store your dye bath, dropping some whole cloves or clove essential oil into the dye bath works well as an antimicrobial.

9. **REVIVING:** If you have stored your dye bath, bring it up to temperature when you're ready to use it, measure the pH, make any necessary modifications, drop in the fibers, and follow the steps above. If there is still color in the dye bath after the first round of dyeing fibers, it can be saved following the above method and used as an exhaust bath, which means adding additional fibers into the dye pot for a second (or third, or fourth) round of dyeing.

ADDING pH MODIFIERS

Modifying the pH of the dye bath can yield different colors. The pH is measured on a scale of 0 to 14, where 7 is considered neutral. Use pH strips or a pH meter to measure the acidity or alkalinity of a dye bath. Mushrooms, unlike other natural dye goods, often need modifications to their pH to coax out their colors. To make an acidic (lower than pH7) dye bath requires the addition of an acid, my preference being household distilled vinegar. To make an alkaline (higher than pH7) dye bath requires the addition of an alkali (base); my preference is soda ash.

Since some mushrooms prefer a dye bath with a modified pH, I do not advise adding mordants directly to the dye pot but instead applying them to the fibers in advance. For example, if alum is added directly to an alkaline (pH9) dye pot it will effectively create a lake, ruining the soluble dye bath and making an insoluble pigment. It is always best practice to rinse the fiber after mordanting to ensure the desired pH of the mushroom dye bath is not impacted.

After fibers have been removed from a dye bath, they can be given an after-bath to shift their colors. For example, fibers dyed in the genus *Hydnellum* could be placed in an alkaline (pH9) after-bath to shift the colors toward blue or green, and any fiber can be placed in an iron after-bath to darken or sadden the colors.

FINISHING THE FIBERS

After removing the fibers from the dye pot, rinse them thoroughly until the water runs clear. Using rinse buckets will help reduce the amount of water used during this process. However, rinsing is not the final step. Be sure to finish your fibers to remove any excess dye and to improve the overall lightfastness and durability of the dyes.

Finishing means bringing dyed fibers up to temperature (see Preparing Fibers for details) in clean water for 10 minutes with a small amount of neutral (pH7) soap. This process swells the fibers, allowing the dye and mordant to penetrate more fully, and removes excess dye from the surface.[4] Be sure to rinse well after finishing.

---

4. Boutrup, Ellis, *The Art and Science of Natural Dyes*, 101.

1/2 tsp/2.5ml

CHAPTER THREE

Pigments

CREATING PIGMENTS

When a pigment is made by precipitating a dye bath onto a substrate to create a solid, insoluble pigment, the process is called a lake. The term *lake* comes from French dyers who discovered how to isolate colored pigments from the insect *kerria lacca*—*lac* for short—from their dye baths. When a mordant and a precipitant are combined in a dye bath, the colorants become suspended and separate from the rest of the solution. The chemical process of making a lake turns the dye into a solid, insoluble pigment. Lake pigments are only ever created from organic materials, such as mushrooms. Inorganic materials, like ochres, are already a solid, insoluble pigment ready to be filtered, separated, and finely ground.

The pH of the mushroom dye bath is important when making a lake. Mushrooms often need a modification to the pH of their dye bath to extract the color. As a result, you may find you need to adjust the ratios of the mordant and precipitant when making a lake to ensure you are capturing all the colorant.

MAKING A LAKE

To make a lake you need to start with a dye bath. The dye bath can either be a fresh one that has not yet been used to dye fibers or it can be an exhaust bath that has already been used to dye fibers. Either option works, but the unused dye bath will yield more colorful pigment. The details below are my preferred methods, but feel free to experiment and tinker with the lake-making process to best serve your creative needs.

1. **PREPARE:** A pot, jar, or bowl can be used to create a lake; however, it needs to have enough room to allow for the precipitate. Lakes have a tendency to bubble up and over small containers, causing a mess and the loss of valuable pigment.

2. **STRAIN:** Remove any solid debris from the dyeing process by straining the dye bath through fine-mesh fiber.

3. **HEAT:** Warm up the dye bath to 85°F/29°C so the minerals can dissolve.

4. **RECIPE:** The rule-of-thumb lake recipe is 10 g of alum and 5 g of soda ash for every 1 liter of water.[1]

5. **MORDANT:** (Alum) This is always added first and poured directly into a warm dye bath. Stir and let dissolve.

6. **PRECIPITANT:** (Soda ash) This is dissolved in hot tap water (110°F/43°C) before adding it to the dye bath with a quick, minimal stir. The more the solution is stirred after it has been added to the dye bath, the harder it will make the pigment, leading to a lot of extra work grinding and mulling.

   When the accumulated pigment settles to the bottom of the jar (a process known as levigation) and the liquid above the settled pigment (aka the supernatant) is clear, the lake solution is ready to be filtered. If the supernatant is not clear, it can be poured off, saved, and re-laked to capture the remaining colorant to make more pigment.

---

1. Boutrup, Ellis, *The Art and Science of Natural Dyes*, 138.

7. **FILTER:** Pour off the supernatant in the sink and filter the pigment through a fine-mesh fiber to strain the contents. This process can take a bit of time, ranging from a few hours to a day or more.

8. **RINSE:** It is important to clean your pigment to remove any excess mineral build-up. The ideal time to rinse the pigment is while it is still a wet paste. My approach is to pour clean water directly over the pigment and strain it. This can be done multiple times to dissolve any remaining minerals.

9. **STORE:** There are two options for storing your pigment: It can be a wet paste or completely dried. Store a wet pigment in a sealed glass jar with a few drops of clove oil as an antimicrobial on a shelf or in a refrigerator. Dry, ground pigments can be stored in a vial or jar with a lid.

MAKING PAINT

Different types of paint can be made by combining lake pigments with a binder, which is a substance that holds the texture of the paint together and creates an adhesion between the paint and the applied surface. Select the binder based on the type of paint you want to make.

The natural world offers many types of binders that can be used to make a variety of different natural paints. The fungi kingdom offers us jelly fungus, which is a type of mushroom that can be used as a binder. To use jelly fungus, incorporate small, chopped pieces of either *Tremella mesenterica*, *Pseudohydnum gelatinosum*, *Guepiniopsis alpina*, or *Exidia glandulosa* directly into a pot with just enough water to cover the mushrooms, and cook them down to form a thicker, gelatinous solution that can be used as binder.

1. **PREPARE**: To make paint or ink you will need to mull the pigment. To do this you will need a mulling tool—I use a special glass muller, but any flat, smooth, hard object like the bottom of a glass, a kraut weight, or even a rock will do the job. You will also need a slab, which must be flat and hard. I use a glass plate that I abraded by mulling it with silicone carbide—the surface must be rough to create enough friction to properly mull the paint.

2. **GRIND:** Take the dried lake pigment and grind it using a (dedicated!) grinder, a mortar and pestle, or both into a finely ground powder.

3. **RECIPE:** I have included my standard recipes for mixing watercolor paint and ink. They can be adjusted based on the amount of pigment being used. Making watercolor paint results in a pan of paint that will dry and can be rehydrated with water using a paintbrush. Making watercolor ink results in a liquid stored in a bottle that can be used with a paintbrush, fountain pen, or dropper.

   A common ratio for mixing paint is one part pigment to one part binder to one part water; however, lake pigments made from mushrooms vary so much I have modified the basic recipe. I substitute some alcohol (gin, vodka, wine, etc.) for the water since it is highly absorptive, but be careful: Adding too much alcohol can create a fluffy pigment that is very hard to mull.

4. **MULL:** Place your pigment on the prepared slab, add the alcohol, and using a palette knife, mix it together. Next, add half the amount of water called for in the recipe, and mix it together. The texture should be similar to toothpaste. If it is not, add the remaining amount of water. If necessary, add more water until it is a soft paste that can easily be mulled.

WATERCOLOR PAINT RECIPE

1 g pigment
0.5 ml alcohol
1.5 ml water
2 ml gum arabic binder

WATERCOLOR INK RECIPE

1 g pigment
1.5 ml alcohol
2.5 ml water
4 ml gum arabic binder

Start mulling in a circular motion. When you start mulling, the mixture will be loud and scratchy; as you continue it will become quiet and smooth. Continue mulling, applying pressure so the mixture becomes less and less granular. Using a palette knife, scrape the mulled mixture back together, evaluate the texture and consistency, and add more water if necessary. Each mushroom lake pigment is a little different, so experiment with the process.

5. **GRIT TEST:** Before adding the binder do a quick grit test. Wipe a bit of mulled pigment onto a piece of paper to evaluate its texture. If it still contains grains of grit, continue mulling. Look for a smooth, supple consistency without any stickiness from the muller. Perform more grit tests as necessary until you are satisfied with the pigment.

6. **MIX:** Using a pipette, add half the gum arabic binder called for in the recipe to the nicely mulled pigment, using a palette knife to mix it together. Start mulling in a circular motion to coat every particle of pigment with binder. If making ink, scrape all the mulled pigment off the slab and into a jar, add the gum arabic into the jar, and stir.

7. **SMUDGE TEST:** Before adding the remaining binder do a quick smudge test. Wipe a bit of pigment onto a piece of paper and run a cloth over the top. If the pigment smears or streaks, add the remaining gum arabic binder and continue mulling. Perform more smudge tests, adding gum arabic binder as necessary, until the paint or ink does not smear or streak. Keep in mind if you add too much binder, your paint won't dry and will end up sticky. If you don't use enough binder, the paint will crack and flake when it is dry.

   At this point, if you would like to improve the flow of your paint or ink, you may add a few drops of glycerin or honey (one or the other). Note that this addition can also decrease the vibrancy of your paint or ink.

Phaeolus schweinitzii
Hydnellum regium

ADDITIONAL RESOURCES

FURTHER READING: A selection of books featuring the use of mushrooms for color, dyes, and pigments.

Dyes from Nature by Riikka Räisänen, Anja Primetta, and Kirsi Niinimäki

Färgsvamper & svampfärgning by Hjördis Lundmark and Hans Marklund

A Heritage of Colour: Natural Dyes Past and Present by Jenny Dean

In Pursuit of Color: From Fungi to Fossil Fuels: Uncovering the Origins of the World's Most Famous Dyes by Lauren MacDonald

Journeys in Natural Dyeing: Techniques for Creating Color at Home by Kristine Vejar and Adrienne Rodriguez

Magic in the Dyepot by Ann Paulsen Harmer

Mushrooms for Color by Miriam Rice; illustrations by Dorothy M. Beebee

Mushrooms for Dyes, Paper, Pigments & Myco-Stix by Miriam C. Rice; illustrations by Dorothy M. Beebee

Natural Dyes: Sources, Tradition, Technology, Science by Dominique Cardon

The Rainbow Beneath My Feet: A Mushroom Dyer's Field Guide by Arleen Rainis Bessette and Alan E. Bessette

Vakre farger fra naturen: farging med sopp og lav by Anna-Elise Torkelsen

SUPPLY HOUSES: Natural material suppliers for *The Mushroom Color Atlas.*

Aurora Silk
aurorasilk.com

Botanical Colors
botanicalcolors.com

Maiwa
maiwa.com

Natural Pigments
naturalpigments.com

INTERNET OFFERINGS: Online learning resources and communities dedicated to the use of mushrooms for color.

European Mushroom Dyers
Facebook Group

Mushroom Color Atlas
mushroomcoloratlas.com

Mushroom & Lichen Dyers United
Facebook Group

Mushrooms for Color
mushroomsforcolor.com

Mycopigments
mycopigments.com

North American Mycological
Association
namyco.org

GLOSSARY

ACID: A chemical compound with a pH below 7 that forms an acidic solution in water.

ADJECTIVE DYE: A type of dye that requires a mordant in order to bind to fiber.

ALKALI: A chemical compound with a pH higher than 7 that forms a basic solution in water.

ALUM: Alum, or aluminum potassium sulfate, is a metallic salt and the most commonly used mordant for protein (animal) fibers.

ALUMINUM ACETATE: A metallic salt, commonly used as a mordant for cellulose (plant-based) fibers in conjunction with a tannin or alkali.

ANTHRAQUINONES: One of the main groups of natural colorants and one of the most stable, creating lightfast, bright colors. An anthraquinone is a water-soluble compound that produces colors from red to yellow and orange. Examples include dermocybin and dermorubin, which are found only in mushrooms and are known for producing good shades of red; emodin, methoxy-flavomannin, and parietin, which produce yellow shades; and fallacinol, which creates orange-reds.

BASIDIOMYCOTA: A large and diverse phylum of club fungi.

BINDER: A substance that holds pigment particles together to form paint.

CAP: The top part of a mushroom.

CINNAMIC ACID: A source of yellow and orange colorants; derivatives include hispidin and hypholomin, which are the result of mushrooms breaking down lignin on live or dead wood.

CITRIC ACID: A colorless organic compound that can be used to make solutions with a pH lower than 7.

COPPER: Copper, or copper acetate, is a dark green crystalline powder that is prepared from a reaction of acetic acid and copper oxide.

DYE: A colored, soluble substance of dissolved molecules, not particles, that bond chemically to a substrate such as fiber.

FASTNESS: The stability and permanence of a dye. A "washfast" dye will not fade when washed and a "lightfast" dye will not fade when exposed to light.

FLAVONOIDS: Derived from the Greek term *flava*, which means "yellow," flavonoids are one of the main groups of natural colorants that contain multiple subgroups such as flavones, flavonols, anthocyanins, and more. A flavonoid is a water-soluble compound that produces colors from orange to yellow. Colorless flavonoids may interact with other flavonoids to form a compound color.

FUNGI: Spore-producing organisms, which include mushrooms, that are neither plants nor animals.

GREVILLINES: A source of yellow and yellow-green colorants found in *Boletus*.

HYMENIUM: The spore-bearing surface of a mushroom, located on the underside of the cap.

IRON: Iron, or ferrous sulfate, can be used as a mordant to affect the color of natural dyes.

LAKE: The process by which a dye (a colored, soluble solution) is precipitated (using an alkali) onto a substrate (a mineral salt mordant) resulting in solid, insoluble particles of color known as a lake pigment. This is the most common process for making a pigment from an organic material.

LEVIGATE: To refine a pigment through sedimentation, grinding, and washing processes.

MISAPPLIED: In mycology, a misapplied name indicates that a mushroom found in the U.S. was originally thought to be the same as a European mushroom; however, either DNA or closer inspection indicated otherwise, and the newly discovered mushroom received its own name. The European name is considered misapplied, even if it persists in common usage.

MODIFIER: A substance that changes the color of a dye bath by shifting the pH.

MORDANT: Derived from the Latin term *mordere*, which means "to bite," a mordant is a chemical that is used to bind a dye to a substrate. Mordants are necessary for dyes that have a very low or no natural affinity to bind with fibers or other substrates. Mordants can modify dye color as well as improve wash- and lightfastness.

MULLING: Refining paint by grinding it on a slab with a muller in a circular motion in order to coat every particle of pigment with a binder.

MUSHROOM: The above-ground, fleshy fruiting body of a fungus.

MYCELIUM: The rootlike structure of a fungus consisting of a mass of branching, thread-like hyphae, often growing underground or inside decaying wood.

MYCORRHIZA: A symbiotic association between a fungus and a plant. They are commonly divided into ectomycorrhizae and endomycorrhizae, where ectomycorrhizal fungi do not penetrate individual cells within the root and endomycorrhizal fungi do penetrate the cells within the root.

NANOMETER: A metric unit of length equal to one billionth of a meter.

pH: Stands for the "potential of hydrogen" and denotes the acidity or alkalinity of a solution. pH is measured on a scale of 0 to 14, where 7 is considered neutral. pH lower than 7 indicates acidity; higher than 7 indicates alkalinity.

PIGMENT: A colored substance which, unlike a dye, is composed of solid, insoluble particles.

POTASSIUM HYDROXIDE: A caustic, inorganic compound with the formula KOH, commonly called caustic potash. It is typically used as a reagent to help identify mushrooms.

PULVINIC ACID: A source of yellow and yellow-green colorants found in boletes.

SAPROPHYTE: A type of fungus that lives on dead or decaying organic matter.

SCOUR: The term for washing and cleaning fibers thoroughly to remove any contaminants before mordanting and dyeing.

SODA ASH: Soda ash, or sodium carbonate, is an alkali base. It is the sodium salt of carbonic acid. It is used as a modifier during the dye process and as a precipitant during the lake-making process.

STIPE: The stem or stalklike feature supporting the cap of a mushroom.

SUBSTANTIVE DYE: A type of dye that does not require a mordant in order to bind to fiber.

SUPERNATANT: The liquid that settles above a solid residue after precipitation.

SYMBIOTIC: Describes the interaction between two different organisms living in close physical association. In the fungi kingdom, the relationship can be mutualistic, where both parties benefit from the interaction, or it can be parasitic, where one party benefits while the other is harmed.

SYNONYM: In mycology, a synonym is a second name for a single mushroom, the result of it being discovered and named by two separate parties. This springs either from a lack of communication or a mistaken belief that the mushrooms were different species.

TANNIN: A group of naturally occurring compounds used in dyeing textiles to improve the wash- and lightfastness of the dyed fiber.

TERPHENYLQUINONES: A subcategory of anthraquinones, terphenylquinones are colorants only found in fungi. Derivatives of terphenylquinones such as polyporic acid, thelephoric acid, and atromentin are the main colorants that produce blue, blue-gray, and violet hues. Flavomentin is a source of orange-yellow colorants.

TIN: Tin, or stannous chloride, can be used as a mordant to modify dye color and improve wash- and lightfastness.

TITANIUM OXALATE: A mordant that can be used with both cellulose and protein fibers.

ACKNOWLEDGMENTS

I am incredibly grateful for the amazing opportunity to have written this book. I am especially thankful to everyone who had a hand in shaping it and for the love, support, and encouragement from my family, friends, and, first and foremost, Brad Johnson. I am immensely grateful for the colorful life we share. The Mushroom Color Atlas website would not exist without your brilliant, beautiful, insightful design and your thoughtful dedication and commitment to bringing it to life. From the bottom of my heart, thank you!

To my editor and champion, Natalie Butterfield, for your vision and belief that there was a book within my work. I am beyond grateful for your invitation to bring it to life. Thank you for shepherding me through the process, providing immensely helpful feedback, and encouraging me each step of the way. A tremendous thank-you to Wynne Au-Yeung for her beautiful book design, attention to every detail, and willingness to work with a spreadsheet! And many thanks to the entire team at Chronicle Books: Kristen Hewitt, Michele Posner, Erin Thacker, Tamar Schwartz, Lesley Bruynesteyn, and everyone else who worked tirelessly to make this book possible and launch it into the world.

The book is graced with stunning illustrations created by the incredibly talented Yuli Gates. It was beyond a pleasure, truly a gift, to collaborate with Yuli. Her enthusiasm, endless research, and overall joy and happiness make these mushrooms radiate an immense beauty, for which I am filled with gratitude.

I couldn't have made this book without John Niekrasz. I am very grateful for his writing prowess and encouragement while expertly editing and wordsmithing the manuscript. Thank you to Charlie Kitchings for the color-filled days in the studio writing, researching, and discovering more about color than we ever imagined. To the esteemed mycologists Dr. Michael Beug, Leah Bendlin, and Danny Miller, who took the time and generously provided insight and knowledge during their review of the species identification and mushroom description text, thank you.

I was lucky to be in the presence of the powerhouse photography team Jack Winegar, Anne Parker, and Dave Fischer, who brought all their talents to my studio for the day. I am thankful for Jack's incredible eye for detail behind the camera, along with his patient and calming presence. I am grateful to Anne for her vision, execution, and beautifully perfected styling and to Dave for all his heavy-lifting assisting.

Danny Rosenberg continues to awe me with his programming genius, making it all look so easy, simple, and effortless. His sense of humor always keeps me on my feet, and I am indebted to his all-around amazing brilliance in helping bring the Mushroom Color Atlas website to life.

A heartfelt thank-you to my parents, who nurtured and instilled a deep love of the natural world, among many other things, in me. To all the generous mycophiles and natural dyers who have helped me on my journey—you mean the world to me! Oregon Mycological Society Dyers (Ellen, Vicky, and Carolyn); Wildcraft Studio School (Chelsea, Rachel, and Kate); Oregon College of Art & Craft Natural Dye Study Group (Judilee, Jo, Isabella, Sally, and Janelle); and the Packwood Mushroom Dye Crew (Alissa, Anna, Erica, Jen, Kristi, Lynda, Marion, Sarah, Stacy, Tess, Thea, and Teddy). Of course, I can't forget my most patient, loyal companion, Pahto, who is always up for a forage, no matter the weather conditions or time of day.

Last, but not least, thank you to the fungi for making life possible on planet Earth. And to all of you, readers, thank you for your enthusiastic, curious interest in using mushrooms for color. Here's to always learning, experimenting, sharing, and discovering all the colors that the powerful fungi kingdom has to offer us.

INDEX

A

B

C